T0298517

# Prediction and Control of Noise and Vibration from Ventilation Systems

This book addresses the prediction and control of noise and vibration in ventilation systems and their psychoacoustic effects on people. The content is based on the authors' research and lecture material on building acoustics and provides insights into the development of prediction methods and control of noise and vibration from ventilation systems, and an assessment of their psychological effects on people.

The basic principles and methods for prediction and control of noise and vibration from ventilation systems are discussed, including the latest developments on flow-generated noise prediction, assessment methods for the performance of vibration isolation, noise control using periodic Helmholtz Resonators, and holistic psychoacoustic assessment of noise from ventilation systems.

Features:

- The book provides insights on noise and vibration in ventilation systems
- Extends into prediction, control, and psychoacoustic assessment methods

This book suits graduate students and engineers specialising in acoustics, noise and vibration control, building services engineering, and other related fields within the built environment.

# Prediction and Control of Noise and Vibration from Ventilation Systems

Cheuk Ming Mak, Kuen Wai Ma, and
Hai Ming Wong

CRC Press
Taylor & Francis Group
Boca Raton London New York

CRC Press is an imprint of the
Taylor & Francis Group, an **informa** business

Cover image: Cheuk Ming Mak, Kuen Wai Ma, and Hai Ming Wong

First edition published 2023
by CRC Press
2385 NW Executive Center Drive, Suite 320, Boca Raton FL 33431

and by CRC Press
4 Park Square, Milton Park, Abingdon, Oxon, OX14 4RN

*CRC Press is an imprint of Taylor & Francis Group, LLC*

© 2024 Cheuk Ming Mak, Kuen Wai Ma, and Hai Ming Wong

*Library of Congress Cataloging-in-Publication Data*
Names: Mak, Cheuk Ming, author. | Ma, Kuen Wai, author. | Wong, Hai Ming, author.
Title: Prediction and control of noise and vibration from ventilation systems /
Cheuk Ming Mak, Kuen Wai Ma, and Hai Ming Wong.
Description: First edition. | Boca Raton : CRC Press, [2024] |
Includes bibliographical references and index.
Identifiers: LCCN 2023004167 | ISBN 9781032061986 (hardback) |
ISBN 9781032062013 (paperback) | ISBN 9781003201168 (ebook)
Subjects: LCSH: Ventilation–Noise. | Noise control.
Classification: LCC TH7681 .M34 2024 | DDC 697.9/2–dc23/eng/20230216
LC record available at https://lccn.loc.gov/2023004167

ISBN: 978-1-032-06198-6 (hbk)
ISBN: 978-1-032-06201-3 (pbk)
ISBN: 978-1-003-20116-8 (ebk)

DOI: 10.1201/9781003201168

Typeset in Sabon
by Newgen Publishing UK

To our families

# Contents

# Preface

This book addresses the prediction and control of noise and vibration in ventilation systems and their psychoacoustic effects on people. The content is based on past research and lecture material of the authors on acoustics. This book is not purported to be a comprehensive book on noise and vibration control for ventilation systems, because there are several handbooks or practical guides for noise and vibration control for ventilation systems. Instead, this book provides insights into the development of prediction methods and control of noise and vibration from ventilation systems, and an assessment of their psychological effects on people.

Summarising all the noise and vibration problems in ventilation systems in one book is a daunting and impossible task. In Chapter 1, we have tried to provide knowledge of building acoustics and psychoacoustics to readers looking for basic knowledge to support their understanding of the rest of the book. Chapter 2 briefly introduces duct-borne sound prediction and control, as there are design guides and handbooks that provide tables and charts for the prediction of transmission noise in air ducts. It also provides a stepping stone to the more complicated duct-borne noise problem discussed in Chapter 3. Chapter 3 begins by highlighting the engineering importance of the prediction of flow-generated noise from ventilation ductwork systems, and then introduces past works on the measurement of flow-generated noise and the development of prediction techniques for flow-generated noise from ventilation systems. Presumably, this is the first reference book to provide such a detailed review and introduction to the prediction methods of flow-generated noise from air ducts, particularly pressure-based prediction techniques. Chapter 4 introduces the fundamental knowledge of vibration isolation and the assessment of the vibration isolation performance of vibrating ventilation equipment. Machine vibrations are complex. In practice, only a simple single-degree-of-freedom undamped lumped-parameter system, with a single contact point, is used to predict the vibration-isolation performance. This chapter introduces a more accurate prediction method, that is the power transmissibility method, based on the overall structure-borne sound power transmission and dynamic characteristics of machines and floor structures. To understand the effect of the acoustic environment on people owing to vibration or sound sources, Chapter 5 discusses psychoacoustic approaches. The psychoacoustics perception scale (PPS), the first psychometric tool designed to quantitatively assess all fundamental perceptual dimensions of sound, is introduced. Objective psychoacoustic metrics and subjective human responses in psychoacoustic prediction are discussed. Case studies and examples are provided to the readers in Chapter 6 to illustrate the prediction of duct transmission noise, flow-generated noise, vibration isolation performance, and human psychoacoustic response discussed in the previous chapters.

The authors hope this book helps inspire readers to get involved in this topic.

*Cheuk Ming Mak*
*Kuen Wai Ma*
*Hai Ming Wong*

# Authors

 **Cheuk Ming Mak** is a professor at the Hong Kong Polytechnic University, specialising in acoustics and building environments. He obtained his bachelor's degree and PhD from the University of Liverpool and earned a diploma in Acoustics and Noise Control from the Institute of Acoustics (UK) after receiving his undergraduate degree. He has been conducting research on noise and vibration from ventilation systems and their effects on people for more than 30 years.

 **Kuen Wai Ma** is currently a postdoctoral fellow at the Hong Kong Polytechnic University. He obtained his PhD degree in psychoacoustics from the University of Hong Kong in 2020 after earning a BSc degree in physics from the Chinese University of Hong Kong. His doctoral and postdoctoral research focused on the characterisation, statistical modelling, and prediction of human perceptual responses to different types of indoor and outdoor environments.

 **Hai Ming Wong** is a clinical professor at the University of Hong Kong, specialising in paediatric dentistry and the effects of noise on health and comfort. She obtained her Doctor of Dental Surgery from Chung Shan Medical University, her Master of Dental Science in paediatric dentistry from the University of Liverpool, and her PhD and advanced diploma in paediatric dentistry from the University of Hong Kong. She has been conducting numerous studies to link the noise problem to increased anxiety.

Chapter 1

# General building acoustics

## 1.1 INTRODUCTION

Noise and sound, which are both vibrations of air particles, are physically the same. Considering human aural perception, noise is typically regarded as an unwanted sound. The energy properties of acoustical environments are commonly characterised by measurements of the sound pressure level (SPL; IEC 801-22-07), which are instantaneous sound pressure fluctuations. Because there is a wide range of detectable sound pressures from the human threshold of hearing (root mean square (RMS) sound pressure of 20 µPa), SPL is measured on a logarithmic scale in terms of decibel (dB). Therefore, SPL cannot be averaged arithmetically. Hence, equivalent continuous sound levels (IEC 801-22-16) are applied to measure the time-averaged sound levels over the elapsed time ($T$) of the measurements. Most acoustic environments can be roughly estimated using several SPL measurements at different locations in the environment.

Environmental noise is formed by sounds at different frequencies. Frequency is the number of vibrations of air in a second; it is measured in Hertz (Hz). The audio frequency range is typically 20–20,000 Hz for any person with a normal hearing. Nonetheless, human hearing sensation depends on the frequency of the sounds. The human hearing system is more sensitive to 500–6,000 Hz than other frequencies in the audio frequency range. A-weighting, the most common frequency weighting in acoustical measurements, is a filter that approximates human ear responses. The A-weighted equivalent sound level ($L_{Aeq,\ T}$), a time-averaged energy level of the A-weighted SPL in a period of $T$, is a good indicator of the general acoustic environment.

## 1.2 FUNDAMENTALS OF ACOUSTICS

Acoustics is the scientific study of the generation, transmission, and reception of acoustic energy in the form of vibrational waves in an elastic medium [1]. Specifically, it deals with the space-time evolution of sound in an elastic

DOI: 10.1201/9781003201168-1

medium. Therefore, the sound pressure of a travelling harmonic wave can be represented by a sinusoidal function of space and time dependence. If sound waves propagate in free space without boundaries, sound reflection will not occur. However, sound waves exhibit phenomena such as multiple sound reflections, interference, and absorption when boundaries exist, such as in fully enclosed or semi-enclosed rooms or air ducts.

## 1.2.1 Building acoustics

Acoustics [2] is concerned with the generation and space-time evolution of small mechanical perturbations in a fluid (sound waves) or solid (elastic waves). An *acoustic wave* allows energy to be transferred from one point to another without any mean displacement of air particles in the medium between the two points. *Sound* is an aural sensation caused by pressure variations in air and is always produced by a source of vibration. For example, the compressor of a condenser system gives rise to a succession of compressions and rarefactions, which radiate out from the source at the speed of sound. Sound is not only responsible for the sensation of hearing but is also governed by analogous physical principles. Noise problems in building acoustics are typically in the audio frequency range. This is referred to as the audio frequency range. Sounds with frequencies below 20 Hz are referred to as *infrasound*. Sounds with a frequency above 20,000 Hz are called *ultrasound*, the study of which is called ultrasonics. Both infrasound and ultrasound are regarded as sounds. The difference between acoustics and optics is that acoustics uses mechanical waves as opposed to electromagnetic ones.

*Noise* is an unwanted sound that has a negative impact on people. However, sound sources are inevitably present in the environment. Therefore, perceptions influenced by the acoustic environment are critical in differentiating desirable and undesirable sounds. Building acoustics [3] involves the science of controlling and predicting sound in buildings and studying the effects of sound on building occupants. This is an important issue because it is related to the indoor environmental quality, human comfort, and satisfaction. A holistic approach consisting of both objective and subjective information of the acoustic environment provides more comprehensive knowledge of noise control in buildings.

## 1.2.2 Energy feature

Sound pressure, sound intensity, and sound power are three physical quantities related to the energy content of the acoustic environment. Both levels of these physical quantities are measured in decibels (dB), which is expressed as the ratio of two values of the quantities on a logarithmic scale. One decibel is 1/10 of Bel (B), named after Alexander Graham Bell. The decibel is more famous in acoustics because 1 dB is the only noticeable difference in sound intensity for the normal human ear. The logarithmic scale corresponds more closely to the wide range of the human hearing response, from the standard threshold of hearing at 1,000 Hz to the threshold of pain at some ten trillion times that intensity. In addition, normal algebraic addition is not generally applicable to the addition/subtraction of levels. In general, an increase of 3 dB represents doubling of the quantities. Detailed definitions of the logarithmic and related quantities can be found in the IEC Standard 60027-3:2002.

Sound pressure, $p(t)$, is a measure of the small pressure perturbation above atmospheric pressure in Pascal (Pa). This quantity is affected by the surrounding environment. It can be heard in the audio frequency range and can be measured using the acoustic equipment. The effective sound pressure of an acoustic measurement is referred to as the RMS value of instantaneous sound pressure over a time interval, $T$, at a location, $p_{RMS, T}$. The minimum sound pressure that can be captured by the human ear is 20 µPa ($2 \times 10^{-5}$ Pa). It was chosen as the reference sound pressure, $p_0$, in the SPL (or $Lp$). The SPL value is defined as follows:

$$SPL = 10\log_{10}\left(\frac{p_{RMS},T}{p_0}\right)^2 = 20\log_{10}\left(\frac{p_{RMS},T}{p_0}\right), \tag{1.1}$$

where $p_{RMS,T} = \sqrt{\frac{1}{T}\int_0^T p^2(t)dt}$.

For an arbitrary periodic acoustic wave $p = p_A \sin \omega t$, the value of $p_{RMS}$ is proportional to the amplitude $p_A$ of the sound pressure of the wave, as follows:

$$p_{RMS,T} = \frac{p_A}{\sqrt{2}} \quad or \quad p^2_{RMS,T} = \frac{p^2_A}{2}$$

Two detector response times (Fast and Slow) are commonly applied in SPL measurements. This indicates how rapidly the detector output signal changes in response to a sudden change in the input signal. Fast has a time constant of $T = 0.125$ s, whereas Slow has a time constant of $T = 1$ s.

Sound power, $W$, is the energy emitted by a sound source per unit time measured in Watts (W). This quantity is an inherent property of a sound source, and is independent of the environment. The reference sound power, $W_0$, was chosen to be $10^{-12}$ W in the sound power level ($L_W$) calculation, as follows:

$$L_W = 10\log_{10}\left(\frac{W}{W_0}\right) \tag{1.2}$$

The sound intensity, $I$, is the acoustic energy passing through a unit area per second in watt per area (W/m²). This quantity changes with the distance from the sound sources. The reference sound intensity, $I_0$, was chosen to be $10^{-12}$ Wm⁻² in the sound intensity level ($L_I$) calculation, as follows:

$$L_I = 10\log_{10}\left(\frac{I}{I_0}\right)$$

(1.3)

### 1.2.2.1 Sound field

If a sound wave is emitted by a source and spherically propagated in a *free-field* (direct field) environment with a homogeneous, steady, and simple fluid or elastic solid without reflection, diffraction, absorption, and dispersion, the relationships of $p_{RMS}$, $I$, and $W$ at a distance $r$ from the source will be

$$I = \frac{W}{\text{Area}} = \frac{W}{4\pi r^2} \quad \text{or} \quad I = \frac{p_{RMS}^2}{Z},$$

(1.4)

where $Z$ denotes the characteristic acoustic impedance of the medium. $Z$ is equal to the density of the medium ($\rho$; kgm⁻³) times the speed of sound ($c$; ms⁻¹). For the air media, $Z$ = 428 kgm⁻²s⁻¹ at 0°C and $Z$ = 413 kgm⁻²s⁻¹ at 0°C. The $L_I$ of a loud sound causes pain to the ear, usually between 115 and 140 dB.

It is necessary to understand the sound fields inside an enclosed space for indoor noise control and prediction of noise and vibration from ventilation systems. The sound field in an enclosed space is strongly affected by the reflective properties of the enclosed surfaces in a room. Owing to the reflective surfaces, multiple reflections occur, and a *diffused-field* (reverberant field) is established in addition to the direct field from the sound source. Therefore, at any point in an enclosure, the overall SPL is a function of the energy contained in the direct and reverberant fields.

*Direct sound* is the sound that travels from the source to the receiver without undergoing reflection at any boundary. In real-life situations, sound sources in an indoor room are not perfect point sources that radiate sound energy equally in all directions. Therefore, a directivity factor is required to describe the angular dependence of sound intensity. After applying a directivity factor, $Q_\theta$, to Eqs. 1.4, the intensity due to the direct sound is given by

$$I_{Dir} = \frac{WQ_\theta}{4\pi r^2}$$

(1.5)

Table 1.1 Directivity factor and index of the source at different positions

| Position of source | Directivity factor, $Q_\theta$ | Directivity Index, $DI_{\theta,}$ [dB] |
|---|---|---|
| Centre of room | 1 | 0 |
| On wall or floor | 2 | +3 |
| Wall/floor junction | 4 | +6 |
| Corner, wall/wall/floor | 8 | +9 |

The SPL due to direct sound, in terms of the $L_W$ of the source, is then given by

$$\text{SPL}_{\text{Dir}} = L_{\text{p,Dir}} = L_W + 10\log_{10}\left(\frac{Q_\theta}{4\pi r^2}\right) \tag{1.6}$$

The value of $Q_\theta$ depends on the acoustic characteristics of the source and its position relative to the surface of the room. If the acoustic characteristics of the source are not known, then it is usually sufficient to assume a unidirectional source and take the value of $Q_\theta$ appropriate to the source position in the room (see Table 1.1).

The directivity index $DI_\theta$ is defined as:

$$DI_\theta = 10\log_{10} Q_\theta \tag{1.7}$$

In the free-field condition, there were no surfaces obstructing sound propagation. However, the sound field in a room travels only a short distance before it strikes the surface. Because all the reflected sound rays strike at least one boundary, the SPL of the reverberant sound depends on the amount of absorption at the boundary materials. The reverberant sound level builds up to a level where the rate of the supply of sound energy from the source is equal to the rate at which it is absorbed at the room surface.

The *reverberant sound pressure level*, $\text{SPL}_{\text{Rev}}$, depends on the $L_W$ of the sound, a quantity called the room constant.

$$R_c = \frac{S\bar{\alpha}}{1-\bar{\alpha}} \tag{1.8}$$

which depends on the total area of the room surface $S$ and the average absorption coefficient of the room surfaces, $\bar{\alpha}$. The $\text{SPL}_{\text{Rev}}$ is also given in terms of the $L_W$ of the source as:

$$\text{SPL}_{\text{Rev}} = L_{\text{p,Rev}} = L_W + 10\log_{10}\left(\frac{4}{R_C}\right) \tag{1.9}$$

The absorption coefficient $\alpha$ represents the random incidence absorption coefficient of a material and is defined as the proportion of energy absorbed by a surface to the incident energy on the surface.

The average absorption coefficient, $\bar{\alpha}$, is the average value over all room surfaces, $S_i$, considering the different areas as follows:

$$\bar{\alpha} = \frac{\sum_{i=1}^{n} S_i \alpha}{\sum_{i=1}^{n} S_i} \tag{1.10}$$

All objects in the room, for example, people and furniture, provide absorption and should be considered when calculating the average absorption coefficient of the room. The absorption per object, multiplied by the total number of objects, was added to the top line of the formula. $S_{total}$ does not change when considering objects in the room.

Hence, the total SPL was determined by the logarithmic addition of the direct and reverberant sound levels in Eqs. 1.6 and 1.9.

$$\text{SPL}_{total} = L_W + 10\log_{10}\left(\frac{Q_\theta}{4\pi r^2} + \frac{4}{R_C}\right) \tag{1.11}$$

The direct sound of a point source falls by 6 dB per doubling of the distance, whereas the reverberant sound is the same at all positions in the room.

### 1.2.3 Spectral feature

#### 1.2.3.1 Octave scale

The sound spectrum (IEC 801-21-15) is the magnitude of the sound frequency components. This is also referred to as the spectral distribution of the acoustic energy. The temporal wave of the sound pressure, $p(t)$, can be transformed by Fourier analysis into the frequency spectrum, $L_p(f)$. If the wave is a periodic sound, then a discrete frequency spectrum is obtained. A continuous frequency spectrum was obtained if the wave was a random sound. As most sound waveforms are random in the time domain, real-life waveforms usually have all components within the frequency range in $L_p(f)$. For vibration studies, the frequency bands of the spectrum were in units of Hz. The spectrum may cover the frequency range of infrasound and ultrasound, as the haptic sense caused by vibration will also be of research interest.

For $L_p(f)$ covering the audio frequency range, frequency analysis is essential to identify the acoustic characteristics of the environment, determine the dominant frequency bands, and hence design appropriate noise controls. The octave or 1/3 octave scale and Bark scale are typically applied for frequency analysis in acoustic and psychoacoustic measurements, respectively. The octave (IEC 801-30-09) scale splits the auditable spectrum into ten equal logarithmic frequency bands. 1/3 octaves further divide the spectrum into 33 bands. The 1/3 octave spectrum is typically used to analyse the spectral content of sounds. An audiometry exam test of hearing ability is also based on the diagnosis of hearing thresholds at the octave or 1/3 octave bands.

Each octave band is named by the centre frequency, $f_c$, which is the geometric mean of the upper frequency limit of an octave band, $f_{cut-off\_up}$, and the lower frequency limit of the band $f_{cut-off\_low}$, and is given as follows:

$$f_c = \sqrt{f_{cut-off\_up} \times f_{cut-off\_low}} \tag{1.12}$$

Because the ratio of $f_{cut-off\_up}$ to $f_{cut-off\_low}$ for each octave band is always 2, $f_{cut-off\_up}$ and $f_{cut-off\_low}$ can be expressed in terms of $f_c$ as follows:

$$f_{cut-off\_up} = \sqrt{2} f_c \quad \text{or} \quad f_{cut-off\_low} = \frac{1}{\sqrt{2}} f_c \tag{1.13}$$

Therefore, the bandwidth of an octave band, $B = \left(\sqrt{2} - \frac{1}{\sqrt{2}}\right) f_c$, is not constant. The octave band of the higher $f_c$ results in a larger $B$ of the band. For a 1/3 octave band, the relationship between $f_{up\_c}$ and $f_{low\_c}$ is as follows:

$$f_{cut-off\_up} = \sqrt[6]{2} f_c \quad \text{or} \quad f_{cut-off\_low} = \frac{1}{\sqrt[6]{2}} f_c \tag{1.14}$$

### 1.2.3.2 Bark scale

The Bark scale, also called the psychoacoustical scale, was proposed by Zwicker in 1961 [4]. The upper frequency limit, the lower frequency limit, and the central frequency of all critical bands (Bark bands) in the Bark scale included the measurement results of a series of experiments on the threshold for complex sounds, considering the masking, the perception of phase, and most often the loudness of complex sounds. Compared with the octave scale, the exact relationship between the frequency limits and central frequency cannot be found for the Bark scale. The entire audio frequency range was divided into 24 Barks. The 24-Bark spectrum is represented by the specific loudness, $N'$, over the critical bands. The comparison tables of the different frequency scales in terms of 1/3 octave, octave, and Bark bands are shown in Figure 1.1.

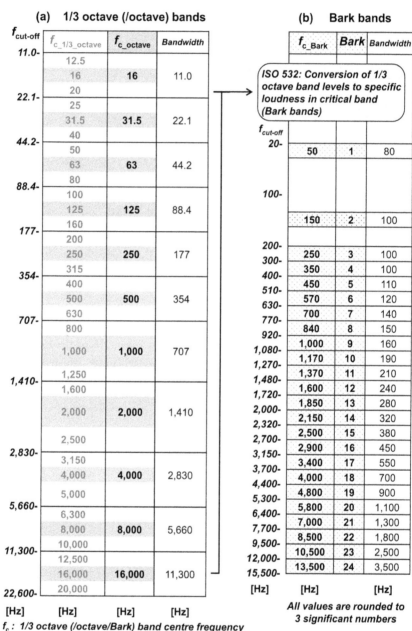

**(a)   1/3 octave (/octave) bands**

| $f_{cut-off}$ | $f_{c\_1/3\_octave}$ | $f_{c\_octave}$ | Bandwidth |
|---|---|---|---|
| 11.0- | | | |
| | 12.5 | | |
| | 16 | 16 | 11.0 |
| | 20 | | |
| 22.1- | | | |
| | 25 | | |
| | 31.5 | 31.5 | 22.1 |
| | 40 | | |
| 44.2- | | | |
| | 50 | | |
| | 63 | 63 | 44.2 |
| | 80 | | |
| 88.4- | | | |
| | 100 | | |
| | 125 | 125 | 88.4 |
| | 160 | | |
| 177- | | | |
| | 200 | | |
| | 250 | 250 | 177 |
| | 315 | | |
| 354- | | | |
| | 400 | | |
| | 500 | 500 | 354 |
| | 630 | | |
| 707- | | | |
| | 800 | | |
| | 1,000 | 1,000 | 707 |
| | 1,250 | | |
| 1,410- | | | |
| | 1,600 | | |
| | 2,000 | 2,000 | 1,410 |
| | 2,500 | | |
| 2,830- | | | |
| | 3,150 | | |
| | 4,000 | 4,000 | 2,830 |
| | 5,000 | | |
| 5,660- | | | |
| | 6,300 | | |
| | 8,000 | 8,000 | 5,660 |
| | 10,000 | | |
| 11,300- | | | |
| | 12,500 | | |
| | 16,000 | 16,000 | 11,300 |
| 22,600- | 20,000 | | |
| [Hz] | [Hz] | [Hz] | [Hz] |

**(b)   Bark bands**

*ISO 532: Conversion of 1/3 octave band levels to specific loudness in critical band (Bark bands)*

| $f_{cut-off}$ | $f_{c\_Bark}$ | Bark | Bandwidth |
|---|---|---|---|
| 20- | | | |
| | 50 | 1 | 80 |
| 100- | | | |
| | 150 | 2 | 100 |
| 200- | | | |
| | 250 | 3 | 100 |
| 300- | | | |
| | 350 | 4 | 100 |
| 400- | | | |
| | 450 | 5 | 110 |
| 510- | | | |
| | 570 | 6 | 120 |
| 630- | | | |
| | 700 | 7 | 140 |
| 770- | | | |
| | 840 | 8 | 150 |
| 920- | | | |
| | 1,000 | 9 | 160 |
| 1,080- | | | |
| | 1,170 | 10 | 190 |
| 1,270- | | | |
| | 1,370 | 11 | 210 |
| 1,480- | | | |
| | 1,600 | 12 | 240 |
| 1,720- | | | |
| | 1,850 | 13 | 280 |
| 2,000- | | | |
| | 2,150 | 14 | 320 |
| 2,320- | | | |
| | 2,500 | 15 | 380 |
| 2,700- | | | |
| | 2,900 | 16 | 450 |
| 3,150- | | | |
| | 3,400 | 17 | 550 |
| 3,700- | | | |
| | 4,000 | 18 | 700 |
| 4,400- | | | |
| | 4,800 | 19 | 900 |
| 5,300- | | | |
| | 5,800 | 20 | 1,100 |
| 6,400- | | | |
| | 7,000 | 21 | 1,300 |
| 7,700- | | | |
| | 8,500 | 22 | 1,800 |
| 9,500- | | | |
| | 10,500 | 23 | 2,500 |
| 12,000- | | | |
| | 13,500 | 24 | 3,500 |
| 15,500- | | | |
| | [Hz] | [Hz] | [Hz] |

*All values are rounded to 3 significant numbers*

$f_c$ : 1/3 octave (/octave/Bark) band centre frequency
$f_{cut-off}$ : Lower(/upper) frequency limit of an octave(/Bark) band

Figure 1.1  Comparison tables of different frequency scale in terms of (a) 1/3 octave (/octave) bands and (b) Bark bands applied in building acoustics.

### 1.2.3.3 Frequency dependency of loudness sensation

It is well known that human hearing thresholds differ at different frequency bands and are related to the age and gender (ISO 7029:2017 [5]) of people. Our ears are frequency-selective, being most sensitive between 500 Hz and 6,000 Hz, compared with our overall hearing range from 20 Hz to 20,000 Hz. The ear is a non-linear device with sensitivity and can tolerate higher loudness levels at low frequencies. Sound loudness is a subjective term that describes the strength of the ear's perception of sound. It is intimately related to intensity but can by no means be considered identical to the intensity. The sound intensity must be factored into by the ear's sensitivity to the frequencies contained in the sound loudness contours (ISO 226:2003 [6]). Based on the equal soundness contours, which indicate the variation for the average human ear, 1,000 Hz is chosen as a standard frequency to define phon. While a pure tone sound is perceived to be as loud as a 60 dB sound at 1,000 Hz, it is believed to have a loudness of 60 phon. The use of phon as a unit of loudness is an improvement over just quoting the level in decibels; however, it is still not a measurement which is directly proportional to loudness. The sone scale was created to provide a linear scale for loudness. One sone loudness corresponds to a level of 40 dB for a 1,000 Hz tone. The empirical formula relating the loudness $N$ (in sones) to the loudness level $L_N$ (in phons) derived from subjective tests on the listener is as follows:

$$N = 2^{\frac{L_N - 40}{10}} \quad \text{or} \quad L_N = 40 + \frac{10}{3}\log_{10} N \quad\quad (1.15)$$

### 1.2.3.4 Frequency weighting

Frequency weighting filters correlate the sound level meter (objective) measurements with subjective human responses. Our subjective response is also loudness dependent; therefore, early sound level meters/analysers included A, B, C, and Z (zero) and linear or flat frequency weightings to consider the responses. The weighting network scales are used in practice to add the loudness contribution from all bands. The most general one is the A-weighting scale, designed to approximate the ear at approximately the 40 phon level (inverted equal-loudness contour). The B- and C-weighting networks are inverted 70 and 100 phon loudness contours, respectively; however, they are less popular currently. All these weighting filters are based on the 1/3 ocatave scale of the SPL measurement data. The unit of the SPL measurement will be changed to dB(A), dB(B), or dB(C) if A-, B-, or C-weighting is applied in the measurement. A-weighting is the most widely used frequency weighting in environmental noise policy, regulations, and guidelines, as it shows a good estimation of the perceived noise level captured by human ears.

## 1.2.4 Temporal feature

The SPL in Eq. 1.1 is the level for a tiny time interval of the detector response time, $T$. For indoor acoustic measurements, the elapsed time, $T_E$, can be up to minutes, hours, and even a day.

### 1.2.4.1 Equivalent indicator

When the noise is not stationary but does not change significantly with time, it is convenient to add up all quadratic pressures $p^2_x(t)$ (i.e. all acoustic energy) over a period of time and provide an average quadratic pressure. Equivalent continuous sound pressure level, $L_{xeq\_TE}$, is therefore designed based on this average quadratic pressure over a given $T_E$. In addition, the value of $L_{xeq\_TE}$ can be calculated from the $N\ (=T_E/T)$ number of the measured SPL during the entire measurement, as follows:

$$L_{xeq\_TE} = 10\log_{10}\left(\frac{1}{T}\frac{\int_0^{T_E} p_x^2(t)\,dt}{p_0^2}\right) = 10\log_{10}\left(\frac{1}{N}\sum_{n=1}^{N}10^{\frac{L_{x_n}}{10}}\right), \qquad (1.16)$$

where $L_{xn}$ is the $n$th instantaneous x-weighted (e.g. A, B, C, and Z) SPL level of the detector response time. For example, sound from ventilation is usually accepted to be steady during operation. The A-weighted equivalent continuous SPL of 30 min, $L_{Aeq\_30mins}$, is a suitable acoustic metric to represent the general acoustic characteristics of the noise.

### 1.2.4.2 Statistical indicator

From a statistical perspective, the value of $L_{xeq\_TE}$ is different from the average SPL $\left(L_{xeq\_TE} \neq \dfrac{1}{N}\sum_{n=1}^{N}L_{x_n}\right)$ and the median SPL $\left(L_{xeq\_TE} \neq L_{x\left(\frac{N}{2}\right)}\right)$.

For the percentile x-weighted SPL, $L_{xN\%}$ is the SPL that exceeds N% of the measurement time $T_E$, where N is usually chosen to be an integer. The statistical indicators $L_{A10}$ and $L_{A90}$ are widely used in measurements to determine the effects of significant noise sources and background noise in the environment, respectively. The maximum x-weighted SPL level ($L_{x\_max}$) and that of minimum ($L_{x\_min}$) are also two statistical indicators that describe the acoustic characteristics of the environment.

### 1.2.4.3 Reverberation time

When the sound source was switched off, the direct sound ceased immediately, and the reverberant sound decayed at a rate that depended on the total absorption $S\bar{\alpha}$ in the room. The reverberation time $RT$ is defined as the time taken for the sound to decay by 60 dB. In practice, a decay of 60 dB is difficult to achieve, and $RT$ is determined by doubling the time for a 30 dB decay e.g., from −5 dB to −35 dB. The Sabine formula developed by Wallace Clement Sabine is an approximate expression for the $RT$ of a room and is as follows:

$$RT = \frac{0.161V}{S\bar{\alpha}}, \tag{1.17}$$

where $S\bar{\alpha}$ is the total absorption of the room and $V$ is the room volume. $RT$ is also the principal parameter for assessing the suitability of the acoustics of a room for the proposed room. In addition, $RT$ values vary with the sound frequency. As the absorption coefficients of room surfaces generally increase with increasing frequency, the $RT$ of a room generally decreases with increasing frequency. The optimum $RT$ quoted can be at the 500 Hz octave band. Although $RT$ is the most common and useful parameter for assessing the acoustical quality of an architectural space, studies [7, 8] have revealed that the energy decay may not be uniform. Other acoustical parameters associated with the early part of energy decay such as early decay time ($EDT$) are also used for assessing the acoustical quality of an architectural space such as a church.

## 1.3 INDOOR ASSESSMENT OF SOUND AND VIBRATION

### 1.3.1 Acoustic criteria

The equivalent continuous SPL $L_{xeq\_TE}$ is a single-value parameter. However, spectral content is important as it provides information about sound quality. Other noise criteria used in assessing the acoustic performance of air-conditioned spaces include noise criterion ($NC$), noise rating ($NR$), and room criterion ($RC$) curves. The properties of noise, including the sound pressure level, loudness, frequency content, tonal components, impulsiveness, and level fluctuations, determine whether the noise is acceptable. Psychoacoustics is related to human auditory response to sound. Psychoacoustic metrics provide a sense of sound quality.

### 1.3.1.1 Noise criterion curves

Noise criterion ($NC$) is one of the earliest designed and widely applicable indoor noise criteria. $NC$ curves (see Figure 1.2) were developed in

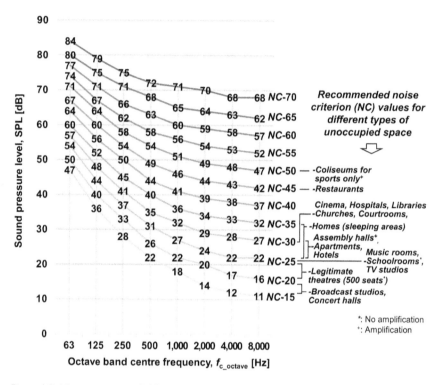

Figure 1.2 Noise criterion (NC) curves and the recommended NC values for different types of unoccupied space.

1957 by Beranek in the U.S.A. [9]. The NC curves include eight octave bands from 63 Hz to 8,000 Hz and the shape of the curves approximates the equal loudness contours developed in the early 1930s by Fletcher and Munson. The NC curves can help return a single NC value of an unoccupied space in buildings from the spectrum of the background SPL of the space. The NC values can serve as a criterion to compare different acoustic environments and check whether the SPL of the acoustic environments is within the acceptable range. The NC value was determined by determining the lowest NC curve that was not reached by the spectrum of the background SPL of the acoustic environment. For example, the octave band spectra of different acoustic environments E1 and E2 are plotted in Figure 1.4(a). The lowest NC curve was not reached by the measured spectra for NC-45. Therefore, environments E1 and E2 had the same NC value as NC-45. This example also demonstrates that NC value is not sufficient to distinguish the spectral features of different acoustic environments. Hence, different indoor noise criteria such as Balanced Noise

Criteria (*NCB*), Preferred Noise Criteria (*PNC*), Room Criteria Mark II (*RC Mark II*) were designed [10] to have a more comprehensive quantification of indoor acoustical environments.

### 1.3.1.2 Noise rating curves

Noise rating (*NR*) curves (see Figure 1.3) were developed in 1962 by Kosten and Van Os. The indoor criteria *NR* is more commonly applied in Europe, while *NC* is more commonly applied in the U.S.A. The *NR* curves

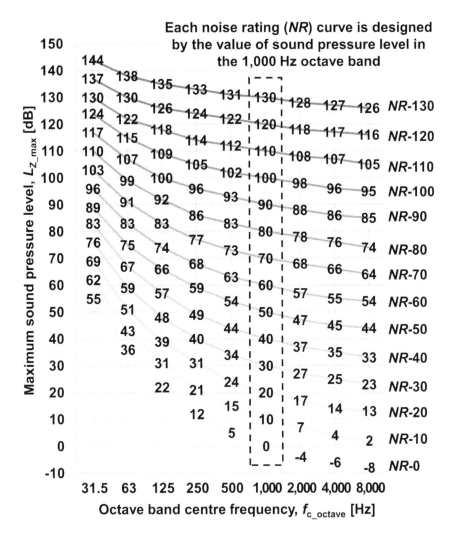

Figure 1.3 Noise rating (NR) curves.

are generated by a formula of the frequency-band dependent constants, a and $b$, which are added to the octave band SPL for $NR$ level, $SPL_{NR} = a + b \times SPL_{octave}$. The naming of each $NR$ curve is designed using the SPL at the 1,000-Hz octave band. Compared with the $NC$ curves, the $NR$ curves include one more octave band (31.5 Hz) and cover a larger range from $NR$ 0 to $NR$ 130. Furthermore, the $NR$ curves are steeper for low-frequency bands, representing a greater tolerance to low-frequency sounds. The determination of the $NR$ value is similar to that of the $NC$ value, to determine the lowest $NR$ curve that is not reached by the SPL spectrum of the environment. The $NC$ value can be used to quantify the acoustic environment and hence as a criterion to assess the acceptance of hearing preservation, speech communication, and annoyance of different environments, as stated in the ISO Recommendation ISO/R 1996-1971 [11]. Using acoustic environments E1 and E2, which have the same $NC$ value as an example (see Figure 1.4(b)), the $NR$ value of E1 ($NR$-44) is different from that of E2 ($NR$-47).

### 1.3.1.3 Room criterion curves

To distinguish the spectral features of different acoustic environments, $RC$ curves were developed. The details of the $RC$ curves can be found in the American National Standard Criteria (ANSI S12.2-2019) for evaluating room noise [12]. The $RC$ curves were also recommended by the American Society of Heating, Refrigerating, and Air-Conditioning Engineers (ASHRAE) on the basis of specifying the design level in terms of a well-balanced spectrum shape. The $RC$ curves included nine octave bands from 16 Hz to 4,000 Hz. The acoustic environment can be regarded as an optimum balance if all the individual octave band SPL of the environment are within ±2 dB of the octave band SPL for the $RC$ level. Moreover, the $RC$ curves not only provide a single $RC$ value but also facilitate the sign of the spectral features of the environment. For example, there are two acoustic environments, E1 and E2, with the same $NC$ value (see Figure 1.4(c)–(d)) and vastly different spectra. In fact, the E1 will be subjectively perceived to be 'rumbly' because it is relatively rich in low frequencies. The E2 will be subjectively perceived to be 'hissy' because it is relatively rich in high frequencies. However, the $NC$ and $NR$ values could not distinguish between the spectral features of E1 and E1. The arithmetic average of the SPL at the 500-, 1,000-, and 2,000-Hz octave bands is the $RC$ value. Therefore, the $RC$ values of E1 and E2 are $RC$-37 and $RC$-44, respectively. Starting from the $RC$ value at the 1,000-Hz octave band, a straight line with a slope of 5 dB per octave band was drawn. Subsequently, two lines (R-line and H-line) were drawn parallel to the straight line. The R-line will be to the left of the 500-Hz band and 5 dB above, and the H-line will be to the right of the 1,000-Hz band and 3 dB above. The sign of the spectral feature of the environment, 'Rumble (R)' or 'Hiss (H)', is given if the SPL spectrum of the environment exceeds the

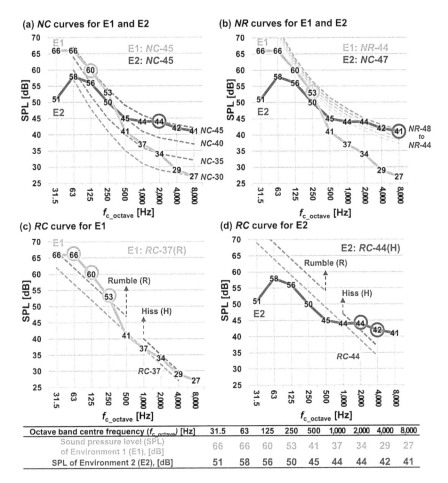

Figure 1.4 Examples of determinations of (a) NC values; (b) NR values; and (c)–(d) RC values of the acoustical environments 1 and 2 (E1 and E2).

R-line or H-line. Therefore, E1 and E2 were determined as $RC$-37 (R) and $RC$-44(H), respectively. If neither the individual octave band SPL of the environment excess the R-line and H-line, the sign of the spectral feature of the environment will be 'Neutral (N)'.

## 1.3.1.4 Psychoacoustics metrics

Psychoacoustics is a science that investigates the objective characteristics and subjective perceptual influence of sound. A psychoacoustic approach that considers both auditory and non-auditory factors provides a higher accuracy in audience perceptual judgements. Psychoacoustic metrics are a series of objective metrics for estimating the actual sensations of sounds based on the psychoacoustical scale (Bark scale) proposed by Eberhard Zwicker in 1961 [4]. Total loudness ($N$), the estimation of the loudness sensation, is the most common psychoacoustic metric in psychoacoustic measurements [4]. The spectral and temporal effects on loudness sensation, the transmission characteristics of human outer and middle ear structures, and the non-linear relationships between sound stimuli and loudness judgements [13] are considered in the $N$ calculation. The international ISO 532 [14] standardises the method of the $N$ calculation method. Moreover, the spectral content of acoustical environments can be characterised by measurements of sharpness ($S$), which is a psychoacoustic metric used to estimate the sharpness sensation by calculating the energy skewness of sounds [4]. The details of the use of the Bark scale in psychoacoustic approaches will be discussed in Chapter 5.

## 1.4 NOISE PROBLEMS

Noise and undesired and unpleasant sounds have always been key issues in environmental management [15]. Environmental awareness of the negative impacts of noise has increased over the last two decades [16]. There is an increasing amount of scientific evidence on the impact of noise [17]. Because the noise level of ventilation sounds is much smaller than the value of 70 dB(A), which causes noise-induced hearing loss in long exposure, psychoacoustic approaches are more important for identifying *psychological impacts* such as acoustic comfort, satisfaction, preference, acceptance, stress, and annoyance from ventilation noise. Other *physiological impacts* [15] such as the risk of high blood pressure, cardiac arrest, tinnitus, and sleep disturbance, and *impacts on performance* [18, 19] such as concentration and memory, are also the focus of researchers in the field.

## 1.5 SOUND AND VIBRATION FROM VENTILATION SYSTEMS

The identification of source noise characteristics is a primary consideration for source noise control. Measurements were required to identify the most significant noise sources in the environment. *Duct-borne sound*, *airborne sound*, and *structure-borne sound* are the three common types of sounds existing in building ventilation systems (see Figure 1.5). In this section, the discussion is primarily focused on engineering practices for noise control of ventilation noise.

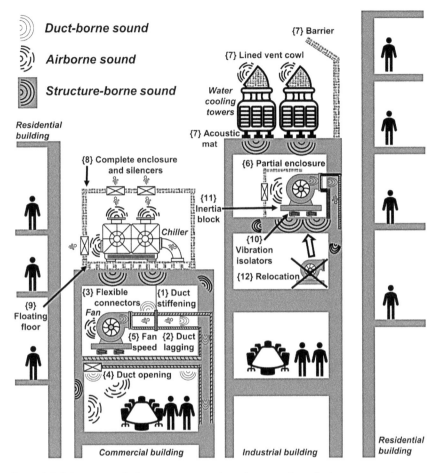

*Figure 1.5* Schematic of the common engineering practices {reference number see Table 1.2} for the noise control of the different types of ventilation noise in buildings.

## 1.6 CONTROL METHODS FOR BUILDING ACOUSTICS

The principal purpose of noise control in building acoustics is to ensure that the acoustic environment of rooms in a building fulfils predetermined acoustic standards and design criteria. Control of sound sources and transmission paths are two major types of engineering practices for ventilation noise in buildings. Mechanical equipment such as air-cooled chillers, water-cooling towers, and fans are well accepted as noise sources in ventilation systems. Therefore, the relocation of noisy equipment, replacement of quieter

*Table 1.2* Recommended engineering practices for the controls of the different type of sounds in ventilation systems

| Type of sounds | Recommended practices | Label in Figure 1.5 |
|---|---|---|
| Duct-borne sound | - Duct stiffening to reduce the duct surface vibrations | {1} |
| | - Duct lagging with sound absorption materials | {2} |
| | - Installation flexible connectors for fan | {3} |
| | - Duct opening away from receivers | {4} |
| Airborne sound | - Fan speed reduction | {5} |
| | - Installation of partial enclosure and silencers | {6} |
| | - Installation of soundproofing barrier, acoustic mat, lined vent cowls | {7} |
| | - Installation of complete enclosure and silencers | {8} |
| | - Installation of floating floor | {9} |
| Structure-borne sound | - Installation of vibration isolators | {10} |
| | - Installation of inertia block | {11} |
| | - Relocation | {12} |

equipment, and routine maintenance of equipment are common engineering practices for controlling sound sources. The basic concept in the control of the sound transmission path involves reducing the propagating sound energy from the noise sources to receivers. Installation of sound absorption materials to equipment surfaces and room surfaces, soundproofing barriers between sources and receivers, partial or complete enclosure of equipment, and silencers to reduce airborne sound are common engineering practices for controlling transmission noise. The common practices for the control of different types of sounds existing in the ventilation systems of buildings are listed and shown in Table 1.2 and Figure 1.5.

### 1.6.1  Controls of duct-borne sound

The sounds of fans in ventilation ductwork systems, as well as the sounds induced by the interaction between solid obstacles and airflow in ducts, can be transmitted between the duct and spaces through the duct surfaces. All these sounds can bother nearby residents at the air outlets of ductwork systems or in spaces involving air ducts. The mechanisms of the different types of duct-borne sounds are discussed in Chapter 2. All of these may cause noise disturbances to nearby residents. Therefore, additional stiffening and composite lagging of sound absorbing materials around the vibrating duct surfaces are effective measures to reduce vibration and induce duct-borne sound at duct surfaces. In addition, the installation of flexible connectors between fans and ducts can help reduce the transmission of fan sounds to a ductwork system. The position change of the duct opening away from the

*Table 1.3* Predicted sound pressure level (SPL) attenuation and percentage drop of
loudness level at different distances from a sound source

| Distance [m] | SPL attenuation [dB] | Drop of Loudness level* [Sone] | Distance [m] | SPL attenuation [dB] | Drop of Loudness level* [Sone] |
|---|---|---|---|---|---|
| 1 | 11 | 0% | 30–33 | 41 | 80% N |
| 2 | 17 | 27% N | 34–37 | 42 | 81% N |
| 3 | 21 | 40% N | 38–42 | 43 | 82% N |
| 4 | 23 | 47% N | 43–47 | 44 | 83% N |
| 5 | 25 | 52% N | 48–53 | 45 | 84% N |
| 6 | 27 | 56% N | 54–59 | 46 | 84% N |
| 7 | 28 | 59% N | 60–66 | 47 | 85% N |
| 8 | 29 | 62% N | 67–75 | 48 | 86% N |
| 9 | 30 | 64% N | 76–84 | 49 | 87% N |
| 10 | 31 | 65% N | 85–94 | 50 | 87% N |
| 11 | 32 | 67% N | 95–106 | 51 | 88% N |
| 12 | 33 | 68% N | 107–118 | 52 | 89% N |
| 13 | 33 | 69% N | 119–133 | 53 | 89% N |
| 14 | 34 | 70% N | 134–149 | 54 | 90% N |
| 15–16 | 35 | 72% N | 150–168 | 55 | 90% N |
| 17–18 | 36 | 73% N | 169–188 | 56 | 91% N |
| 19–21 | 37 | 75% N | 189–211 | 57 | 91% N |
| 22–23 | 38 | 76% N | 212–237 | 58 | 92% N |
| 24–26 | 39 | 77% N | 238–266 | 59 | 92% N |
| 27–29 | 40 | 78% N | 267–298 | 60 | 93% N |

*Notes:* The sound source is assumed to be a point source. * N is the reference value of the psychoacoustic metric (total loudness) measured 1 m from the sound source. The reverberant sound is assumed to be zero.

receivers can also increase the sound transmission path to reduce duct-borne sounds from the ventilation ductwork systems to receivers. The predicted SPL attenuation and percentage drop of the loudness level at different distances from a point sound source are listed in Table 1.3. The predictions of SPL attenuation are based on the inverse square distance law (see Eq. 1.6) and the approximation of the psychoacoustic metric $N$ in the inverse power 0.46 of the distance from a noise source as experimented and proposed by Ma et al., respectively [20].

## 1.6.2 Controls of airborne sound

Airborne sound can be generated from the flow turbulence of rotating fan blades, water spraying and falling in water-cooling towers, and compressor operation of air-cooled chillers in refrigeration cycles in ventilation duct-work systems of buildings. For fan sounds, the reduction in rotational fan speed can help control the SPL of the airborne sound during off-peak hours.

A partial enclosure is usually applied to a stationary noise source that requires space for air supply and/or exhaust or for accessibility. The partial enclosure can be a composite of one or more partition(s), wall(s), barrier(s), or panel(s) to block the transmission path of airborne sounds from a sound source to receivers. The transmission path of sounds can be further increased by placing a barrier closer to the sound source or using a higher barrier. The soundproof performance of the partial enclosure can be increased if the surface(s) facing the sound source are lined with sound absorption materials, and there is a good seal between the surfaces. Silencers can also be installed together with a partial enclosure at the air intake and/ or discharge point(s) of a noise source to absorb the generated airborne sounds. For water-cooling towers, the installation of an acoustic mat and lined vent cowl can help reduce water falling and spraying sounds from the operation. The lined vent cowl can divert the direction of the sound transmission path away from the receiver.

If the sound sources are unavoidably surrounded by residents in all directions, a complete enclosure is considered. The sound reduction index, SRI, is a basic measure of the sound insulation of a partition, and the formula for SRI measurement according to ISO 140-3:1995 [21] is as follows:

$$\text{SRI} = \text{SPL}_{\text{source}} - \text{SPL}_{\text{receiver}} + 10 \times \log_{10} \frac{S}{A}, \tag{1.18}$$

where $\text{SPL}_{\text{source}}$ is the average SPL in the source room, $\text{SPL}_{\text{receiver}}$ is the average SPL in the receiver room, $S$ is the area of the testing material, and $A$ is the equivalent absorption area in the receiver room. The *SRI* of different typical materials according to ISO 15712-1 [22] are shown in Figure 1.6. A complete enclosure is similar to a partial enclosure; all the surface(s) facing a sound source should be lined with sound absorption materials to achieve better performance. Functional and sufficient heat or air exchange is required to prevent equipment overheating. Therefore, the installation of silencers together with a complete enclosure at the air intake and discharge positions is significant for absorbing airborne sounds but allows air exchange from the enclosed equipment to the surroundings. If impact sounds can be generated from the operation of the equipment, the floating floor can be an efficient control measure to isolate the sounds transmitted to receivers in the rooms below the equipment. An acoustic floating floor is a resilient layer suspended above existing structural slabs of buildings. The enclosed air spaces inside the floating floor can help absorb the sounds emitted from the equipment. It can also serve as a protective layer for the existing structural slabs. Therefore, careful defect checking should be performed at the installation stage of the floating floor. Defects are difficult to repair because heavy equipment may be placed above the floating floor.

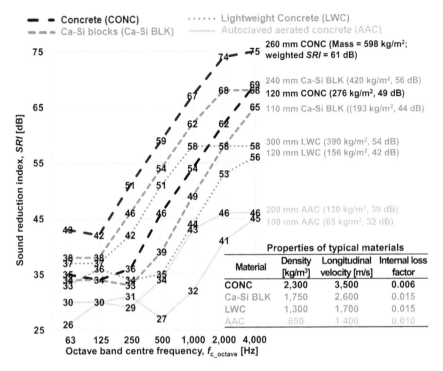

*Figure 1.6* Sound reduction index of the different typical materials according to ISO 15712-1.

Source: From Ref. [22].

## 1.6.3 Controls of structure-borne sound

The transmission of structure-borne sound occurs at the physical contact points between the sound sources and building structure. Therefore, an acoustic floating floor can not only reduce the airborne sound but also minimise the structure-borne sound and vibration generated from the equipment. A floating floor is useful for absorbing sounds and preventing them from arriving at the building structure. Consequently, a more pleasant acoustic environment can be provided to the rooms directly below the equipment. Vibrations generated from the operations of fans, air-cooled chillers, and water-cooling towers can be transmitted and may generate a breakout sound to the different rooms in the building through the building structure. Vibration isolators are a choice other than floating floors to reduce structure-borne sound and vibration transmission from the supporting points of the equipment to the building structure. Metal springs and rubber mounts are two major types of vibration isolators. The efficiency of vibration isolation is critically affected by the selection of the isolator type. The force transmissibility, isolation efficiency, and power transmissibility of the different types of vibration isolators are discussed in Chapter 4. In addition, engineers should consider the high-frequency sound generated from the metallic contacts of metal springs and the relatively short lifetime of rubber mounts. Inertia blocks, which are usually large and heavy concrete blocks, are often installed between the vibrating equipment and vibration isolators to increase the stability and lower the centre of gravity of the ventilation systems. If possible, the relocation of the noisy equipment away from the rooms with receivers also controls the sounds from ventilation systems.

### 1.6.4 Psychoacoustic approach in noise control

The main idea of psychoacoustic techniques is understanding how individuals perceive, experience, and/or understand the acoustic environment. Compared with traditional noise management, psychoacoustic approaches emphasise that environmental sounds are regarded as resources (not wastes); loudness is not the only perception of sounds. The responses to different sound sources need to be differentiated. Noise level reduction is not equal to the improvement of the acoustical environment, and measurements of psychoacoustic metrics other than SPL are required. Differentiating the preferences for different sound sources is important. *Noise masking* with preferable sounds is a possible solution for modifying the public environment. Increased opportunities for exposure to natural sounds, particularly water sounds or music, can promote the general sound quality of the environment.

## 1.7 PREDICTION METHODS FOR BUILDING ACOUSTICS

In the early planning stage of ventilation systems, actual acoustic information of the building may not be available at this stage. Therefore, noise prediction is extremely important for the design of noise controls for ventilation systems [23]. Three prediction methods are widely used in room acoustics: wave-based acoustic, geometrical acoustics, and diffusion equation methods. The application of these methods for the *duct-borne sound, airborne sound, and structure-borne sound* will be further discussed in Chapters 2, 3, and 4, respectively.

### 1.7.1 Wave-based acoustical methods

In wave-based acoustic methods, acoustic waves are assumed to propagate in a homogeneous, steady, isotropic, and perfect elastic medium with boundaries. The wave equations of the acoustic characteristics are completely defined by the functions of space and time. The assumptions of the conservation of mass, momentum, and energy in systems are described by the continuity equation, Euler's equation, and state equation, respectively. These equations can be further simplified by neglecting higher-order terms if the fluctuations of particle velocity, acoustic pressure, and medium density are assumed to be extremely small time-varying perturbations compared to their equilibrium values. The linear time-independent acoustic wave equation is then obtained as follows:

$$\nabla^2 p = \frac{1}{c^2}\frac{\partial^2 p}{\partial t^2},$$ (1.19)

where $p$ is the difference between the instantaneous values of the total and atmospheric pressures. *Finite difference* and *finite element methods* are two common numerical methods used to approximate sound wave propagation in building rooms. This method is rapid and efficient for obtaining the energy features of rooms with simple shapes, well-defined boundary conditions, and dimensions smaller than the wavelength of the wave equation. Currently, numerical simulations are required to not only consider the energy feature but also the frequency feature of the acoustic environment. There is a challenge in terms of the computational power for more complicated situations and frequency-dependent acoustic characteristics.

### 1.7.1.1 Sound wave propagation in air ducts

This section is concerned with sound wave propagation in air ducts. The propagation of sound waves is affected by the reflective properties of fairly rigid surfaces in the air ducts. The effects of the duct walls on the propagation of sound waves in air ducts are briefly outlined. Wave-based acoustical methods can be not only applied to sound wave propagation in building rooms but also to sound wave propagation in air ducts.

Assuming a simple harmonic sound source in a rectangular duct of finite duct length and cross-sectional dimensions a and b, the interior of the duct is the region where 0<x<a, 0<y<b, and –∞<z<∞. In addition, we assumed that the duct walls were rigid.

Assuming a simple harmonic motion of a particular frequency, we have a three-dimension Helmholtz equation obtained from Eq. 1.19

$$\nabla^2 \varphi + k^2 \varphi = 0,$$ (1.20)

where $k = \dfrac{\omega}{c}$ and $\nabla^2 = \left( \dfrac{\partial^2}{\partial x^2} + \dfrac{\partial^2}{\partial y^2} + \dfrac{\partial^2}{\partial z^2} \right)$.

Solving this equation using the method of separation of variables, the following solutions are obtained:

$$\varphi = X(x)Y(y)Z(z), \tag{1.21}$$

where

$$X(x) = A_x \cos k_x x + B_x \sin k_x x \tag{1.22}$$

$$Y(y) = A_y \cos k_y y + B_y \sin k_y y \tag{1.23}$$

$$Z(z) = A_z e^{jk_z z} + B_z e^{-jk_z z}, \tag{1.24}$$

where
$A_x$, $A_y$, $A_z$, $B_x$, $B_y$, and $B_z$ are arbitrary constants, and the wavenumbers must satisfy the condition

$$k^2 = k_x^2 + k_y^2 + k_z^2. \tag{1.25}$$

Boundary Conditions:
The normal particle velocity is zero at the walls $x = 0$ and $x = a$, and the normal particle velocity is zero at walls $y = 0$ and $y = b$.
   Hence, we obtain

$$X(x) = \sum_m A_{x\_m} \cos \frac{m\pi x}{a} \tag{1.26}$$

similarly,

$$Y(y) = \sum_n A_{y\_n} \cos \frac{n\pi y}{b} \tag{1.27}$$

and

$$\pm k_{z\_mn} = \pm \sqrt{\left\{ k^2 - \left( \frac{m\pi}{a} \right)^2 - \left( \frac{n\pi}{b} \right)^2 \right\}} \tag{1.28}$$

Therefore, the complete solution is:

$$\varphi(x,y,z)e^{j\omega t} = \sum_{m,n=0}^{\infty} \cos\left(\frac{m\pi x}{a}\right)\cos\left(\frac{n\pi y}{b}\right)\left(A_{mn}e^{i\left(\omega t-k_{z\_mn}z\right)}\right.$$
$$\left. + B_{mn}e^{i\left(\omega t+k_{z\_mn}z\right)}\right),$$

(1.29)

where

$A_{x\_m}$, $B_{y\_n}$, $A_{mn}$, and $B_{mn}$ are arbitrary constants.

It can be observed from Eq. 1.29 that the complete solution consists of an exponential function multiplied by the modal shape function. *Duct modes* are definite patterns of variation in the amplitude over the cross-section of the air duct because of the interference among the multiple reflections from the duct walls.

For both $m$, $n = 0$ (the zero order mode), we can observe from Eq. 1.28, $k_{z\_mn} = k$.

Plane waves propagate along the duct because there is no dependence on $x$ and $y$.

For the first and higher order modes,

$$\pm k_{z\_mn} = \pm\sqrt{\left\{\left(\frac{\omega}{c}\right)^2 - \left(\frac{m\pi}{a}\right)^2 + \left(\frac{n\pi}{b}\right)^2\right\}}$$

(1.30)

When $\left(\frac{\omega}{c}\right)^2 > \left(\frac{m\pi}{a}\right)^2 + \left(\frac{n\pi}{b}\right)^2$, $k_{z\_mn}$ is real. The sound wave propagates in the $\pm z$ direction, and the *eigenfunctions* given by Eq. 1.29 correspond to *propagating modes*.

When $\left(\frac{\omega}{c}\right)^2 < \left(\frac{m\pi}{a}\right)^2 + \left(\frac{n\pi}{b}\right)^2_{mn}$, $k_{z\_mn}$ is imaginary. A standing wave decays exponentially with $z$ and the *eigenfunction* given by Eq. 1.29 is the *decaying mode*.

The limiting value of $\frac{\omega}{c}$ for which $k_{z\_mn}$ remains real is given by $\left(\frac{\omega}{c}\right)^2 = \left(\frac{m\pi}{a}\right)^2 + \left(\frac{n\pi}{b}\right)^2$, which defines the *cut-on frequency*.

$$\left(\frac{2\pi f_0}{c}\right)^2 = \left(\frac{m\pi}{a}\right)^2 + \left(\frac{n\pi}{b}\right)^2$$

(1.31)

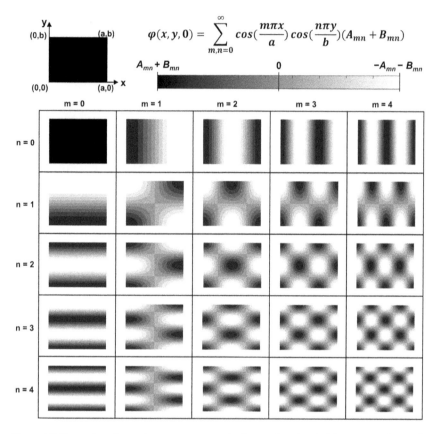

Figure 1.7 Examples of sound wave propagates in the ±z direction in (a) a *propagating mode* and (b) a *decaying mode* at t = 0 s.

The cut-on frequency, $f_0$, is therefore given by:

$$f_0 = \left(\frac{c}{2}\right)\sqrt{\left(\frac{m}{a}\right)^2 + \left(\frac{n}{b}\right)^2} \qquad (1.32)$$

For frequencies below the cut-on frequency of the first transverse duct mode, all modes are evanescent (*cut-off*), and only plane waves are propagated, as shown in Figure 1.7. For simplicity, the magnitudes ($A_{mn}$ and $B_{mn}$) of the waves propagating in the +z and −z directions were assumed to be the same, as shown in Figure 1.7.

For frequencies above the cut-on frequency of the first transverse or higher-order duct modes, the modes are propagated, as shown in Figure 1.8.

**(a) Real Part of** $\varphi(0,0,z)$ **in propagating mode, where** $\frac{\omega}{c} > (\frac{m\pi}{a})^2 + (\frac{n\pi}{b})^2$

$\longleftarrow B_{mn}e^{jk_{z,mn}z} \quad A_{mn}e^{-jk_{z,mn}z} \longrightarrow$

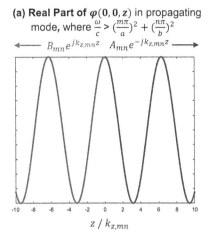

$z / k_{z,mn}$

**(b)** $\varphi(0,0,z)$ **in decaying mode, where** $\frac{\omega}{c} < (\frac{m\pi}{a})^2 + (\frac{n\pi}{b})^2$

$\longleftarrow B_{mn}e^{k_{z,mn}z} \quad A_{mn}e^{-k_{z,mn}z} \longrightarrow$

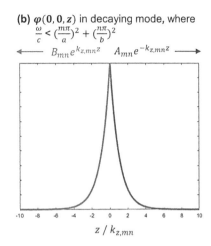

$z / k_{z,mn}$

*Figure 1.8* Examples of the modal sharp functions in different duct modes (m, n) for sound propagation in a rigid air duct of a dimension of a × b m at t = 0 s.

## 1.7.2 Geometrical acoustics methods

Geometrical acoustic methods are also called ray-acoustic methods. They involve the approximate theory; high-frequency sound in air is imagined as travelling in a straight line directly from the sound source to the receiver. The image source, ray tracing, and beam tracing methods can be classified as geometric-acoustic methods. The principal task of geometrical acoustics involves determining the trajectories of sound rays. Rays have the simplest form in a homogeneous medium, where they are straight lines.

## 1.7.3 Diffusion equation methods

A diffusion equation models the sound field in enclosures by diffusely reflecting boundaries. It has been successfully applied to predict room acoustic parameters, such as SPL, in room acoustics simulations. *Finite difference method* was used to obtain approximate solutions under the assumption of sufficiently scattered sound that requires completely diffusive boundary conditions. The *absorption and scattering coefficients* of the boundaries were also considered in the diffusion equation models.

## 1.7.4 Machine-learning methods

Compared with traditional acoustic approaches, the results of machine-learned methods are data-driven models instead of explicit equations of acoustic and non-acoustic variables. Any type of data can be the input and output of machine learning methods. The training of the models was based on datasets of the designed input and output variables. For example, training data can consist of acoustic and psychoacoustic metrics as input variables and psychological and physiological noise impacts as output variables. The features of complex real-life situations are learned and stored in weight matrices inside the machine-learned models. Machine-learned models can predict a new situation based on previous training datasets; however, the internal structures of the models are complicated for humans to comprehend. This is why machine-learned methods are also called black-box methods. Nevertheless, the rapid development of machine learning methods has significant potential for noise control in the future.

## 1.8 SUMMARY

In this chapter, basic knowledge and fundamentals of building acoustics are provided. An overview of the prediction and control methods used in building acoustics was also presented. In Chapter 2, more discussion about duct-borne sound prediction and control is provided. The primary noise, breakout noise, and secondary noise are introduced and discussed. The focus will be on prediction methods for primary and breakout noises, and updated research development of duct-borne sound control, such as the use of periodic Helmholtz Resonators. In Chapter 3, the basic concept of aerodynamic sound is introduced. A review of past studies on fundamental work, data collection, and prediction methods for flow-generated noise is provided. An updated development of prediction methods is discussed. In Chapter 4, the fundamentals of vibrations are introduced. The conventional method for the assessment of vibration isolation is explained, and a recent development of methods for the assessment of vibration isolation is discussed. In Chapter 5, the noise indices commonly used for the assessment of noise from ventilation systems are introduced. The recent development of a holistic psychoacoustic prediction method for noise from ventilation systems is introduced. In Chapter 6, case studies and examples are provided to illustrate the prediction of duct transmission noise, flow-generated noise, vibration isolation performance, and human psychoacoustic response, as discussed in previous chapters.

## References

1. L.E. Kinsler, A.R. Frey, Austin, A.B. Coppens, and J.V. Sanders, 2000 *Fundamentals of acoustics*, 4th Edition. John Wiley & Sons: New York.
2. P. Filippi, A. Bergassoli, D. Habault, and J. Lefebvre. Acoustics: *Basic physics, theory, and methods*, 1998, Elsevier: Amsterdam.

3. C.M. Mak, 2015, *Building and Environment* 94, p.751, Special issue on building acoustics and noise control.
4. E. Zwicker and H. Fastl *Psychoacoustics: Facts and models*, 1990, Springer Science & Business Media: Berlin/Heidelberg.
5. International Standardization Organization, 2000 *ISO 7029:2017 Acoustics – Statistical distribution of hearing thresholds related to age and gender*. International Standardization Organization: Geneva, Switzerland.
6. International Standardization Organization, 1987 *ISO 226:2003 Acoustics – Normal equal-loudness-level contours*. International Standardization Organization: Geneva, Switzerland.
7. M. Mehta, J. Johnson, and J. Rocafort, 1999 *Architectural acoustics: Principles and design*. Prentice Hall: Hoboken, NJ.
8. Y. Chu and C.M. Mak, 2009, *Applied Acoustics* 70, pp.579–587, Early energy decays in two churches in Hong Kong.
9. L.L. Beranek, 1957, *Noise Control* 3, 1957, pp.19–27, Revised criteria for noise in buildings.
10. L.M. Wang and E.E. Bowden, 2003 *Architectural engineering 2003: Building integration solutions*, pp.1–4, Performance review of indoor noise criteria. American Society of Civil Engineers (ASCE); Reston, Virginia.
11. International Standardization Organization, 1971 *ISO/R 1996:1971 Assessment of noise with respect to community response*. International Standardization Organization: Geneva, Switzerland.
12. American National Standards Institute, 1995 *ANSI/ASA S12.2-2019 Criteria for evaluating room noise*. American National Standards Institute: U.S.A.
13. S.S. Stevens, 1955 *The Journal of the Acoustical Society of America* 27, pp.815–829, The measurement of loudness.
14. International Standardization Organization, 1975 *ISO 532-1:2017 Acoustics–Method for calculating loudness level*. International Standardization Organization, Geneva, Switzerland.
15. B. Berglund, T. Lindvall, D.H. Schwela, and World Health Organization, 1999 Guidelines for community noise.
16. World Health Organization. 2011 Regional Office for Europe, *Burden of disease from environmental noise: Quantification of healthy life years lost in Europe*.
17. W. Passchier-Vermeer and W.F. Passchier, 2000 *Environmental Health Perspectives* 108, pp.123–131, Noise exposure and public health.
18. C.M. Mak and Y. Lui, 2012 *Building Services Engineering Research and Technology* 33, pp.339–345, The effect of sound on office productivity.
19. S.X. Kang, C.M. Mak, D.Y. Ou, and Y.Y. Zhang, 2022 *Applied Acoustics* 201, p.109096, The effect of room acoustic quality levels on work performance and perceptions in open-plan offices: A laboratory study.
20. K.W. Ma, C.M. Mak, and H.M. Wong, 2020 *Applied Acoustics* 168, p.107450, Acoustical measurements and prediction of psychoacoustic metrics with spatial variation.
21. International Standardization Organization, 1978 *ISO 140-3:1995 Acoustics – Measurement of sound insulation in buildings and of building elements – Part 3: Laboratory measurements of airborne sound insulation*

*of building elements*. International Standardization Organization: Geneva, Switzerland.

22. International Standardization Organization, 2005 *ISO 15712-1:2005 Building acoustics – Estimation of acoustic performance of buildings from the performance of elements – Part 1: Airborne sound insulation between rooms*. International Standardization Organization: Geneva, Switzerland.

23. C.M. Mak and Z. Wang, 2015 *Building and Environment* 91, pp.118–126, Recent advances in building acoustics: An overview of prediction methods and their applications.

# Chapter 2

# Duct-borne sound prediction and control

## 2.1 INTRODUCTION

The main purpose of a ductwork system is to provide a suitable amount of fresh or conditioned air to an indoor environment. To achieve a good indoor environmental quality, acoustic requirements are as important as thermal requirements in the design of ventilation and air-conditioning systems. To accomplish the acoustic requirements and design criteria, we need to acknowledge the different types of duct-borne noise problems and their generation, transmission, and prediction. In general, three types of noise problems may be encountered in a typical ventilation ductwork system. These are the *primary noise*, *breakout noise*, and *secondary noise*.

Primary noise is also known as *transmission noise*. The source of primary noise in ventilation ductwork systems is fans, which provide air movement and overcomes the pressure drops across the ductwork systems. Because there is much well-established primary noise information [1, 2], it is not a problem for the ventilation system designer to determine how much of the sound generated by fans is transmitted through ductwork systems. In addition, there are well-established methods for selecting attenuators to control the fan noise transmitted through ducted systems. Hence, problems arising from primary noise sources are relatively rare.

Breakout noise is the sound that leaves the system at points other than duct terminations. Therefore, it is more difficult to solve problems arising from the 'breakout' of noise through the walls of the ductwork along which it is passing. In general, this may be due to fan noise, aerodynamic noise from the interaction of airflow, and duct fittings or airborne noise arising from the turbulent airflow interacting with duct walls.

Secondary noise is also known as *regenerated* or *flow-generated noise* and is a type of *aerodynamic noise*. Generally, aerodynamic noise in ducts is produced by the interaction between solid obstacles and high-speed or unsteady airflow. In ventilation system design, it is common to have duct corners, changing duct cross-sections, branch points, dampers, and terminal diffusers. Therefore, with these geometrical discontinuities, the production

DOI: 10.1201/9781003201168-2

of turbulence is unavoidable. The noise produced in this manner in ventilation systems is known as regenerated noise because it is generated on the quiet side of the primary attenuators of the ventilation system. With the advent of high-velocity ventilation systems, and as building materials and structures are becoming lighter, the prediction and possible minimisation of regenerated noise is of great importance. It is known that all in-duct elements generate flow noise in a ventilation system; even an attenuator that is supposed to reduce noise, also generates noise. Consequently, it is very important to predict the amount of flow noise generated during the design stage. A more detailed discussion of flow-generated noise is provided in Chapter 3.

## 2.2 PRIMARY NOISE

To estimate the attenuation of transmission noise for practical engineering purposes, we need to consider the sound power levels of the fan fed into the ductwork system, the acoustic energy being attenuated by the system (such as attenuation provided by silencers), and room corrections.

### 2.2.1 Sound power levels of the fan

A ventilation fan is a device operated by the aerodynamic action of a rotating impeller to continuously drive air through the ductwork system to the indoor environment. Fan noise is caused by mechanical noise from motors and bearings, aerodynamic noise due to rotational noise produced by the interaction between the impeller blades and the surrounding air, and vortex noise due to the turbulent flow across the impeller blades. Different types of fans affect the operating and noise characteristics to suit the different needs of applications. There are two major types of fans.

1.  Centrifugal fans that propel air centrifugally (Figure 2.1(a))
2.  Axial flow fans that propel air axially (Figure 2.1(b))

In general, the noise spectrum of an axial flow fan has a peak at a higher frequency than that of a centrifugal fan of similar duty.

### 2.2.2 Ductwork system attenuation

The original acoustic energy fed into the system by the fan was reduced as it passed down the ductwork system. The attenuation of this acoustic energy by each component of the ductwork system is briefly discussed herein. Tables and charts for ductwork system attenuation can be found in practical design guides or handbooks [1–4].

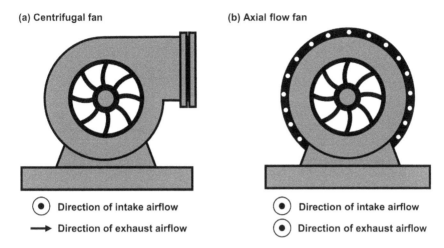

(a) Centrifugal fan

(b) Axial flow fan

● Direction of intake airflow

➡ Direction of exhaust airflow

● Direction of intake airflow

● Direction of exhaust airflow

*Figure 2.1* Schematic diagrams of (a) a centrifugal fan and (b) an axial flow fan.

### 2.2.2.1 Attenuation per unit duct length

Noise from the fan is transmitted and attenuated along the duct since the duct walls are not perfectly rigid. Sound attenuation is due to excitation of the duct wall vibration; molecular air absorption inside the duct can be ignored, as shown in Figure 2.2. Part of the excitation energy is used to overcome the internal damping of the duct material. Consequently, it was converted into heat. The remaining excitation energy is radiated from the duct as the breakout noise. In practice, it is usually assumed that sound power produced by the fan is continuously and linearly reduced along the plain duct length such that the reduction of sound power levels by the duct of constant duct cross-sectional area can be expressed as 'attenuation per metre run' (i.e., dB/m). The ductwork can be circular or rectangular in cross section. The circular duct tends to be stiffer than the rectangular duct of the same cross-sectional area because more acoustic energy can pass along the circular duct. The circular duct exhibits less in-duct attenuation than the rectangular duct at low to medium frequencies.

### 2.2.2.2 Attenuation due to bends and duct fittings

The attenuation of sound energy is obtained by the partial reflection of medium- and high-frequency sound back towards the source at bends or by the corresponding geometry changes where airflow changes direction or is expanded/restricted.

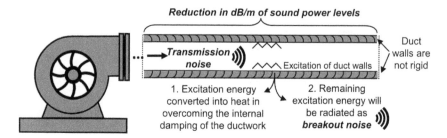

*Figure 2.2* Sound attenuation along the duct length.

### 2.2.2.3 Attenuation at ductwork contractions and expansions

As mentioned earlier, sound attenuation can be obtained at the points of ductwork contractions or expansions. The mechanism involved is similar to that of a bend, in which medium- and high-frequency sounds are reflected along the duct. The amount of sound attenuation obtained is a function of the steepness of the area change (i.e. the more acute, the higher the attenuation) and the ratio of the cross-sectional areas of the ductwork.

### 2.2.2.4 Attenuation due to branches

The total acoustic energy at a branch was divided between the main and branch ducts. It is assumed that this division of energy is directly proportional to the division of airflow, and there is no sound absorption or reflection at a branch.

### 2.2.2.5 Attenuation due to duct terminations

When the sound energy reaches the end of the duct, it goes into the volume of the ventilated room. Owing to the extreme increase in volume (or change in acoustic impedance), acoustic energy loss occurs. These losses indicate that some of the acoustic energy was reflected in the duct space and their amount was frequency dependent. In other words, the losses depend on the wavelength of the sound and the size of the duct. This acoustic attenuation at duct termination is called 'end reflection loss', which were particularly prominent at low frequencies and for smaller areas of termination.

## 2.2.3 Control of sound in the ventilated room

The sound pressure level at any point in a room consists of direct and reverberant sounds.

$$L_{p,\text{Total}} = 10\log_{10}\left(10^{\frac{L_{p,\text{Dir}}}{10}} + 10^{\frac{L_{p,\text{Rev}}}{10}}\right) \tag{2.1}$$

Direct sound is influenced by the sound power emerging from a single outlet, the distance between the source (the duct outlet) and the receiver (a person inside the room) and directivity factors (surface and source). The reverberant sound is influenced by all the duct outlets in the room and is dependent on the room volume and total absorption available from the room surface or contents.

## 2.2.4 Noise control methods for ventilation ductwork system

In addition to sound control within the room, noise control methods that can be used to reduce duct-borne sound transmission include the use of a plenum chamber, package attenuators or silencers, duct liners or semi-active noise control, and active noise control. Although passive noise control methods, such as duct liners and silencers, seem to consume more material resources than semi-active and active systems, the latter requires sensors, controllers, resonators or loudspeakers, and electricity. According to a study conducted by Mak et al. [5] that combined the analytic hierarchy process and life cycle analysis to obtain a holistic view of the system performance in terms of human, economic and environmental impacts, it was found that the most sustainable design was duct liners, followed by silencers, active noise control, and semi-active noise control. This is probably the reason why duct liners and dissipative silencers are the most common noise control methods used in the noise control of ventilation ductwork systems. However, traditional methods, including dissipative and passive reactive silencers, suffer from some serious drawbacks, despite their wide use in duct noise control. The dissipative silencer performs well at medium to high frequencies but fails to be effective at low frequencies because of its high-impedance characteristics. The accumulation of dust and bacterial breeding in porous sound absorption materials is a notable issue related to public health. Passive reactive silencers, Helmholtz Resonators (HRs), and expansion chambers are typical examples that exhibit stable noise-attenuation performance and can be tuned conveniently. Nevertheless, the volume of the expansion chamber must be sufficiently large to handle low-frequency noises. The presence of HR offers a solution to low-frequency noise control; however, it is regarded as a narrow-band noise attenuator that is only effective at its resonance frequency over a relatively narrow frequency range. Over the years, several investigators have attempted to devise a method for controlling duct-borne low-frequency noise over a wide working frequency range. An array of differently tuned HRs or

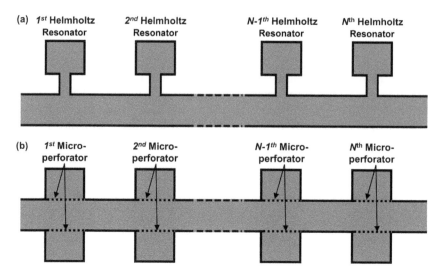

*Figure 2.3* Schematic diagrams of (a) a periodic array of *N* Helmholtz Resonators and (b) periodic array of *N* micro-perforated mufflers.

micro-perforated tube mufflers was studied by Mak's research team [6–12] to reduce low-frequency broadband noise, as shown in Figure 2.3. The use of an array of periodic HRs mounted on the duct can provide a much broader low-frequency noise attenuation band, owing to the coupling effects of the Bragg reflection and resonance of the HRs.

## 2.2.5 Summary

Noise is generated by the fan and transmitted through the ductwork. The noise emerging from duct termination was finally discharged into the ventilated room. A summary of the system noise attenuation is shown in Figure 2.4.

## 2.3 BREAKOUT NOISE

Breakout noise may radiate from a long duct run when there is a high level of internal sound fields or internal flow turbulence; a simple predictive equation is available in the CIBSE guide [2] for the estimation of the sound power radiated from the length of the duct.

$$L_{w,F} = L_{w,D} - R + 10\log_{10}\left(S_d/A_d\right),$$

(2.2)

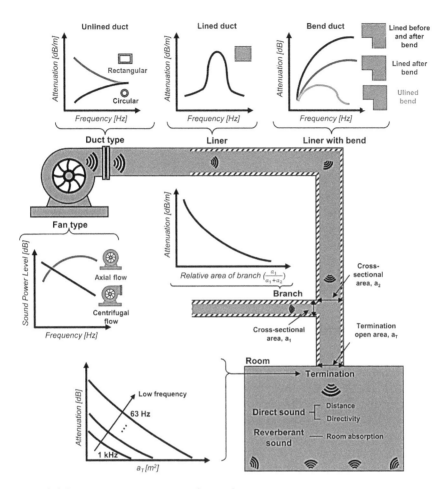

*Figure 2.4* Primary noise transmission from a fan to a room.

where $L_{w,F}$ is the sound power radiating from the duct (W), $L_{w,D}$ is the sound power in the duct (W), $R$ is the sound reduction index of the duct wall material, $S_d$ is the surface area of the duct wall (i.e. duct perimeter × duct length) (m²) and $A_d$ is the cross-sectional area of the duct (m²).

This is an equation for the estimation of turbulence-induced vibration. Because the appropriate values of the parameters in the equation are not easily obtained, this equation has limited practical application. It is difficult to accurately predict breakout noise because it may result from high levels of transmission noise, flow-generated noise, or duct wall vibration because of duct internal sound fields or internal flow turbulence.

To estimate the levels of breakout noise due to the vibration of the duct wall caused by the interaction of the turbulent flow and the in-duct elements, a prediction method based on the static pressure drops across the in-duct elements was proposed [13]. The total sound power levels, $L_{w,Total}$, of the breakout noise due to sound- and turbulence-induced vibrations can be estimated by,

$$L_{w,Total} = 10\log_{10}\left( 10^{\frac{L_{w,A}}{10}} + 10^{\frac{L_{w,F}}{10}} \right),\tag{2.3}$$

where $L_{w,A}$ is the radiated sound power level of the breakout noise due to sound-induced vibration and $L_{w,F}$ is the radiated sound power level of the breakout noise due to turbulence-induced vibration. $L_{w,F}$ can be obtained using Eq. 2.2, and $L_{w,D}$ can be calculated using Nelson and Morfey's equations [14] (discussed in Chapter 3).

Noise breakout may be due to (1) high levels of transmission noise; (2) flow-generated noise; and (3) turbulent airflow interacting with duct walls, causing the duct walls to vibrate and radiate airborne noise.

## 2.4 SECONDARY NOISE

Extensive research on regenerated noise has been conducted, and has had a strong emphasis on jet engine noise. The airflow speeds involved in this situation are generally close to the speed of sound. Compared with the considerable effort in the jet noise field, little work has been conducted on the problem of noise generation by turbulent flow in ventilation duct systems.

In the past, some investigators worked on the collection of data concerning noise generation using various duct elements. However, they worked on a very limited range of duct elements and sizes because of the need for expensive acoustic and aerodynamic facilities for conventional measurements. Other researchers have attempted to develop general techniques for predicting flow-generated noise in ventilation systems. The techniques are based on an empirical relationship between the loss of static pressure caused by the in-duct flow elements and the sound power generated.

The spectrum of the sound power generated was predicted using a generalised spectrum based on a Strouhal number ($St$), for which it is necessary to identify a representative dimension. The simple spoilers used in the experiments of these workers could be accomplished by simple observation. These techniques were developed from a study of the sound generated by simple in-duct elements, such as strip spoilers and orifice plates. Real in-duct elements, such as bends in ventilation systems, will have very dissimilar sizes and shapes compared to these simple in-duct elements. Therefore, it is difficult to determine a representative dimension for various types and sizes of in-duct elements found in ventilation systems. Unless a simple method for determining the representative dimension can be devised, a generalised prediction technique for regenerated noise in a ventilation system cannot be developed.

As most investigators were unable to determine the overall turbulence field in the vicinity of the in-duct elements, they assumed that the fluctuating drag force acting on the flow spoiler was proportional to the steady-state drag force. Because the sound power level of the regenerated noise is determined by the strength of the fluctuating drag force, they assumed that the sound power level of the regenerated noise could be related to the steady-state drag force. Hence, the sound power level of the regenerated noise could be related to the pressure drop across the in-duct elements. Nevertheless, the origin of the regenerated noise is the interaction between turbulence and in-duct spoilers. It would be preferable if the sound power level could be directly related to the turbulence field.

Recently, considerable advances have been made in *computational fluid dynamics* (CFD) such that several commercial codes, which can be used to predict the behaviour of moving air, are now available. In principle, these CFD codes can be used to determine the fine details of airflow turbulence, which can then be used to investigate the noise generated in the ventilation system.

There are many potential advantages to using CFD instead of experimental measurements. First, the most important advantage of computer simulations is their low cost. Second, computational simulation can usually be performed with remarkable speed compared to an equivalent real experiment. Third, a computer solution to a problem provides a considerable amount of detailed information, but no experimental study can be expected to measure the distribution of all variables over the entire domain. Fourth, realistic conditions can be simulated in CFD; for example, various sizes of in-duct elements and ducts with different airflow rates can be modelled.

However, there is a disadvantage to using CFD. The validity of a CFD solution depends on both the mathematical model and numerical method employed, but the experimental investigation, in contrast, observes reality.

Secondary noise is also called *regenerated noise* or *flow-generated noise*. More discussion is presented in Chapter 3. A summary of flow-generated noise is presented in Figure 2.5.

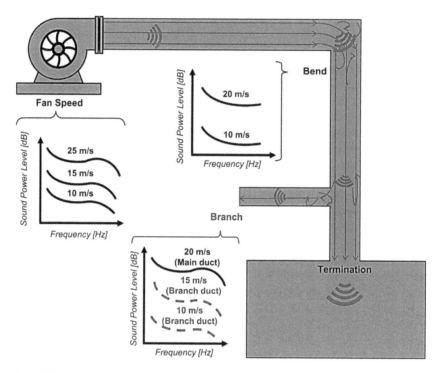

*Figure 2.5* Secondary noise in a ductwork system.

## 2.5 SUMMARY

This chapter has introduced three types of noise problems that may be encountered in a typical ventilation ductwork system. Next chapter will focus on the discussion of prediction methods for flow-generated noise from ventilation systems.

## References

1. American Society of Heating, Refrigerating and Air-Conditioning Engineers, 2019 *ASHRAE Handbook*, HVAC Applications SI Edition. American Society of Heating, Refrigerating and Air-Conditioning Engineers (ASHRAE): Peachtree Corners, Georgia.
2. Chartered Institution of Buildings Services Engineers (London), *CIBSE Guide B5–Noise and Vibration Control for HVAC*, 2002, pp.8–9. Chartered Institution of Building Services Engineering (CIBSE): London, United Kingdom.
3. I. Sharland, 1991 Woods of Colchester Limited, *Woods practical guide to noise control*, Fifth Edition. Flkt Woods Limited: Colchester, UK.

4. Sound Research Laboratories Ltd., 1988 *Noise control in building services.* Pergamon Press: Oxford, UK.
5. C.M. Mak, W.M. To, T.Y. Tai, and Y. Yun, 2015 *Indoor and Built Environment* 24, pp.128–137, Sustainable noise control system design for building ventilation systems.
6. X. Wang and C.M. Mak, 2012 *The Journal of the Acoustical Society of America* 131, pp.1172–1182, Wave propagation in a duct with a periodic Helmholtz resonators array.
7. X. Wang and C.M. Mak, 2012 *The Journal of the Acoustical Society of America* 131, pp.EL316–EL322, Acoustic performance of a duct loaded with identical resonators.
8. Y. Yun and C.M. Mak, 2013 *Building Services Engineering Research and Technology* 34, pp.195–201, The theoretical fundamentals of an adaptive active control using periodic Helmholtz resonators for duct-borne transmission noise in ventilation systems.
9. C.Z. Cai and C.M. Mak, 2016 *The Journal of the Acoustical Society of America* 140, pp.EL471–EL477, Noise control zone for a periodic ducted Helmholtz resonator system.
10. C.Z. Cai and C.M. Mak, 2018 *Applied Acoustics* 134, pp.119–124, Hybrid noise control in a duct using a periodic dual Helmholtz resonator array.
11. C.Z. Cai, C.M. Mak, and X. Wang, 2017 *Applied Acoustics* 122, pp.8–15, Noise attenuation performance improvement by adding Helmholtz resonators on the periodic ducted Helmholtz resonator system.
12. X.F. Shi and C.M. Mak, 2017 *Applied Acoustics* 115, pp.15–22, Sound attenuation of a periodic array of micro-perforated tube mufflers.
13. N. Han and C.M. Mak, 2007 *Technical Acoustics* 25(6), pp. 653–657. Estimation of breakout sound power level due to turbulence caused by an in-duct element.
14. P.A. Nelson and C.L. Morfey, 1981 *Journal of Sound and Vibration* 79, pp.263–289, Aerodynamic sound production in low speed flow ducts.

# Chapter 3

# Prediction methods for flow-generated noise from ventilation systems

## 3.1 INTRODUCTION

In the previous chapter, we discussed the three types of noise problems commonly encountered in ventilation ductwork systems. In this chapter, we focus on the development of prediction methods for flow-generated noise from ventilation systems. *Flow-generated noise*, also known as *regenerated noise* or *secondary noise*, is the noise generated on the quiet side of primary attenuators because of the interaction between the turbulent air flow and duct discontinuities or flow spoilers, as shown in Figure 3.1. In other words, it is the additional noise generated by unsteady flow in the wake of an obstruction in a flow duct.

At long distances from a noise source, such as a fan in a flow duct, flow-generated noise from in-duct elements can be severe. Splitter attenuators or dissipative silencers, which are commonly used in ducts to absorb the noise produced by ventilation system supply fans, can generate noise as well as absorb it. Hence, controlling the flow-generated noise of duct fittings is of utmost engineering importance.

If flow-generated noise could be predicted at the design stage, then it would not be a problem as a suitable remedy could be prescribed. In fact, the failure to accurately predict noise levels, especially in high-speed (i.e., air flow velocities above 10 m/s) ventilation flow ducts, can make it impossible to solve problems that arise after commissioning because there is usually no space available.

Ventilation system designers usually attempt to predict flow-generated noise using the procedures set out in publications such as the *ASHRAE Handbook* [1] or *CIBSE Guide* [2]. The information contained in these publications were drawn from the published works of different researchers. However, these researchers usually consider a limited range of in-duct components and duct sizes. Thus, applying their results to systems with configurations considerably different from those on which the original measurements were made can produce questionable results. In short, existing design methods are inadequate. Willson and Iqbal [3] have claimed that

DOI: 10.1201/9781003201168-3

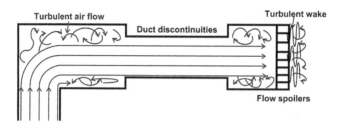

*Figure 3.1* A schematic of flow-generated noise in a ventilation system.

the standard techniques for predicting flow-generated noise due to in-duct elements, found in the *ASHRAE Handbook* and *CIBSE Guide* significantly underestimate the noise levels likely to be encountered in practical systems.

## 3.2 THE CONCEPT OF 'AERODYNAMIC NOISE'

Because the mechanism of flow-noise generation is directly related to aerodynamics, the flow-generated noise from ventilation systems is a form of 'aerodynamic noise'. It is therefore important to understand the concept of aerodynamic noise and the different types of aerodynamic sound sources before the previous work on flow-generated noise is reviewed.

Aerodynamic noise is noise generated as a by-product of an unsteady (turbulent) airflow, or noise produced without (vibration) the participation of solid boundaries. Its frequency spectrum extends from the sub-audible to the ultrasonic, encompassing a wide range of frequencies of comparable intensity.

Aerodynamic noise can be produced by air jets, boundary layers, vortices and wakes, edge tones, and related phenomena. The theory of aerodynamic sound generation was modelled after the stress system in a fluid acting on fluid elements. This leads to the condensation and dilation of the fluid elements, causing them to act as acoustic radiators [4]. Numerous factors contribute to stress, including pressure and velocity field fluctuations, shear forces in the flow, fluid viscosity, and external body forces.

As previously stated, aerodynamic noise also comprises ultrasonic noise. It cannot be 'heard' but could have detrimental effects on human beings or animals, and thus require investigation by medical researchers. The discussion here will be restricted to the low flow speeds encountered in ventilation systems that result in the generation of low-frequency sound, and for which current aerodynamic theory provides three fundamental mechanisms of noise generation. Figure 3.2 (after Gordon [5]) shows a schematic representation of the typical locations and descriptions of the three noise-generation mechanisms for a terminal duct carrying air.

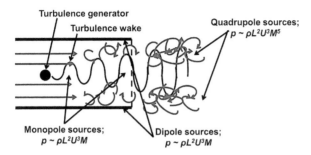

*Figure 3.2* A schematic of the three noise-generation mechanisms for a terminal duct carrying air.

## 3.2.1 Aerodynamic monopole

Aerodynamic monopole radiation occurs when mass or heat is introduced into a fluid at a non-steady rate. A monopole source is similar to a pulsating sphere (Figure 3.3 after Howe and Baumann [6]). The acoustic pressure-pulse fronts are always in phase, thereby producing a spherical directivity pattern.

When monopole sound is generated by unsteady flow velocities, the dimensional relationship between the radiated sound power and important parameters that produce it is:

$$W_{monopole} \propto \frac{\rho L^2 U^4}{c} = \rho L^2 U^3 M \qquad (3.1)$$

where
$W_{monopole}$ = radiated sound power, W;
$\rho$ = mean gas density, kg/m³;
$c$ = speed of sound in gas, m/s;
$U$ = flow velocity, m/s;
$L$ = length scale; and
$M$ = Mach number which is equal to $U/c$, dimensionless.

It follows that, for a monopole source, the intensity of sound generation is proportional to the fourth power of the mean flow velocity.

Typical monopole sources include pulse jets (where high-speed air is periodically ejected through a nozzle), sirens (where a steady airflow is periodically chopped), and propellers at zero pitch (where air is periodically displaced each time a blade passes a given point).

In ventilation systems, the necessary perturbation of the net mass flow through the duct system, which results in a monopole source, could arise

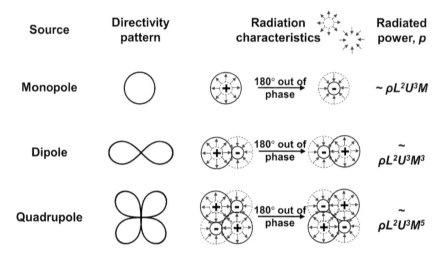

*Figure 3.3* Typical sources of aerodynamic noise and their dimensional properties in the fluid of uniform mean density.

from a fluctuating stall in the air-moving equipment or, less likely, from the net compressibility of the duct air flow.

### 3.2.2 Aerodynamic dipole

When turbulent flow interacts with a solid surface, or when turbulence (in the form of a wake) is generated by a solid surface, changes occur in the momentum field around the spoiler. Such changes are time-dependent; require fluctuating forces; and must, in general, exist on the solid surface. Any such forces may be resolved into 'lift' and 'drag' components, and the sound generated is represented by an acoustic dipole (Figure 3.3), the axis of which is aligned with that of the fluctuating force. The dipole can be considered as two monopoles separated by a small distance and pulsating out-of-phase with each other. Most sound generation mechanisms in air-conditioning systems can be represented by dipole characteristics.

In the flow system shown in Figure 3.1, the presence of the spoiler indicates the location of a dipole source. The figure also indicates the possibility that the turbulent wake from the spoiler impinges on the duct wall at the exit plane and generates surface noise.

If the dipole-source distribution is random, then the sound spectrum is of a broadband nature. A well-ordered distribution, on the other hand, results in a discrete sound spectrum. This is what happens with rotational noise in fans.

The dimensional dependence of the radiated dipole sound power is as follows:

$$W_{\text{dipole}} \propto \frac{\rho L^2 U^6}{c^3} = \rho L^2 U^3 M^3 \qquad (3.2)$$

This indicates that the radiation intensity from the dipoles obeys the sixth power of the velocity law. Dipole radiation is a factor of $M^2$, which differs from that of a monopole. At subsonic speeds (where $M < 1$), the dipole is less efficient than the monopole. Because of the pressure cancellation that occurs in the plane normal to the dipole axis, the contributions yield a directivity pattern with a figure-of-eight shape that is strongest in the direction of the dipole axis, and weakest at right angles to the axis.

### 3.2.3 Aerodynamic quadrupole

In airflow systems, turbulent mixing can occur in the absence of solid surfaces. Examples of this are found in the induction of airstream mixing and boundary layer turbulence. When air leaves the duct, it enters a region of zero flow velocity. The high-velocity gradients at the interface of the moving and stationary media produce shear forces in the fluids, which generate turbulent mixing and noise. The noise sources are spatially distributed along and across the regions of mixing. Lighthill [7, 8] proposed that a quadrupole represents a model of the acoustic source. The quadrupole can be considered as a combination of two dipoles aligned in opposite or parallel directions, as shown in Figure 3.3.

This is the acoustic quadrupole phenomenon generally referred to as 'jet' noise. This noise source dominates high-speed subsonic, turbulent air jets. The strength of the quadrupole source is large in regions where both turbulence and mean-velocity gradients are high, such as in the highly turbulent mixing layer of a jet. Sources of noise generated by turbulence may also be found within ducts. These are associated with local high-velocity gradients, as in the spoiler wake in Figure 3.3.

The dimensional dependence of radiated quadrupole sound power is as follows:

$$W_{\text{quadrupole}} \propto \frac{\rho L^2 U^8}{5} = \rho L^2 U^3 M^3 \qquad (3.3)$$

The radiated sound power for this source type is proportional to the eighth power of the flow velocity. The quadrupole efficiency differs by factor $M^2$ from that of the dipole. At subsonic speeds ($M < 1$), the quadrupole radiation efficiency is lower than that of the dipole because of the double cancellation effect, as shown in the fourth column of Figure 3.3.

### 3.2.4 Dominant noise sources

The radiation efficiencies of monopole, dipole, and quadrupole sources decrease in subsonic flows. However, the dependence of their radiated sound powers on flow speeds exhibit the opposite trend, i.e., the total radiated sound power varies as the fourth, sixth, and eighth power of the flow speed for monopole, dipole, and quadrupole sources, respectively. Thus, despite the inherently low efficiency, radiation from a quadrupole source may dominate that from other sources at a high flow speed. The quadrupole radiation for a jet engine with a high exhaust speed usually predominates, although other internal sources, such as rough burning (predominantly a monopole source) or compressor noise (predominantly a dipole source), are present and contribute to the total noise.

The constant of proportionality for each type of source may have different values depending on the sound generation process. Thus, the constant for a singing-wire phenomenon is different from that for an edge-tone phenomenon, although both result from aerodynamic dipole radiation. The proportionality relations in Eqs. 3.1, 3.2, and 3.3 are useful in estimating the effects of parameter changes on the radiated sound power.

### 3.2.5 Additional ideas on aerodynamic sources

Hardy [9] has shown that by assuming a simple spherical diffusion source, it is possible to deal with phenomena such as air outlets under pressure (monopoles), fan noise (dipoles), and jet noise (quadrupoles).

By considering the sound radiation from unsteady forces resulting from turbulence in the oncoming flow and from the shedding of vortices from a flow spoiler, Curle [10] showed that under certain conditions, the applied forces on a fluid correspond to acoustic dipoles. Several authors [11, 12] have shown that the dipole (surface) sound radiated by an infinite rigid flat plate vanishes, and the radiation field degenerates into a quadrupole type. In considering a finite plate, however, where the flow extends over the edges, the pressure differential across the plate is relieved by the flow around the edges, generating a sound similar to that from 'pressure dipoles' – a type of radiation that cannot exist elsewhere on the plate [13].

The analysis of the noise generated by flow spoilers usually assumes that the spoilers are small compared with the sound wavelength at the frequency in question. This ensures that the flow field in the neighbourhood of the spoiler may be treated as incompressible. Davies and Ffowcs–Williams [14] have noted that when a sound source is confined within an infinitely long duct with a small diameter compared to the wavelength, the character of the problem changes from three- to one-dimensional. This change affects both the acoustic impedance of the surroundings and the amount of sound power radiated by the source. In addition, the dependence of the sound power on the frequency of the source changes by a factor $f^2$.

At low frequencies, the monopole, dipole, and quadrupole sources confined in a small-diameter duct would have sound power outputs increasing as $u^2$, $u^4$, and $u^6$ instead of the typical $u^4$, $u^6$, and $u^8$ characteristics of these sources in an infinite fluid, respectively. As the frequency of the sound source increases, the sound power radiated by the source in the duct approaches that in the free field.

## 3.3 PREVIOUS WORKS ON FLOW-GENERATED NOISE FROM VENTILATION SYSTEMS

The purpose of this review is to indicate the amount of effort spent so far on the measurement and development of prediction techniques for flow-generated noise.

Several researchers have contributed to the development of prediction techniques for flow-generated noise in ventilation systems. Prior to actually developing such techniques, some researchers established the theory or basis of the mechanism of aerodynamic noise generation [15].

Before 1950, the production of sound by airflow in the presence of rigid bodies was of interest, especially in connection with musical instruments. Theoretical and experimental investigations of the intensity of the sound produced by fluid flow were performed.

It is known that turbulent motion produces noise directly related to the fluctuating pressure distribution. In 1950, Batchelor [16] obtained pressure correlations by considering the pressure fluctuations caused by the turbulent motion of a viscous incompressible fluid; however, he did not consider how to estimate the sound generated by turbulence.

In the early 1950s, a new impetus for research on aerodynamic noise was provided by the theory of Lighthill [7, 8] which predicted the sound intensity produced by fluid motion. This became the classical theory of aerodynamic noise generation that describes the mechanism by which the free turbulence in the mixing region of a jet exhaust radiates sound. He demonstrated that the sound field is produced by the static distribution of acoustic quadrupoles, and that the intensity of the noise generated increased according to the eighth power of the exhaust velocity. The basis of this approach was the comparison with the governing equations of the density fluctuations of real fluid, which are appropriate for a uniform acoustic medium at rest. The difference between the two sets of equations was considered as the effect of a fluctuating external force field (externally applied fluctuating stresses) acting on the uniform acoustic medium at rest, hence radiating sound according to the ordinary laws of acoustics.

In 1952, Proudman [17] analysed the generation of noise by homogeneous isotropic turbulence on the basis of Lighthill's theory.

In 1955, Curle [10] extended Lighthill's general theory of aerodynamic sound by incorporating the influence of solid boundaries on the sound field. First, the reflection and diffraction of sound waves occur at the solid boundaries. Second, there is a resultant dipole field at the solid boundaries, which are the limits of Lighthill's quadrupole distribution. These effects are shown to be equivalent to a dipole distribution.

In 1963, Powell [18] attempted to describe the aerodynamic sound generated by the movement of vortices, or vorticity, in an unsteady fluid flow. His theory is particularly well-structured to estimate sound from flows described in terms of vorticity.

In 1967, Davies and Ffowcs–Williams [14] addressed the problem of estimating the sound field generated by a limited turbulence region in an infinitely long, straight, and hard-walled pipe. The acoustic power in the pipe was considered and calculated for two types of turbulent motion. For the first type, the eddies large to such an extent that the motion is completely correlated across the pipe, and all the sound is in the form of a plane wave propagating in the axial direction. The second type of motion is a statistically slow varying flow with a small eddy correlation length compared with the cross-sectional pipe dimension.

After the basis of the aerodynamic sound generation theory was established, numerous theoretical and experimental studies were conducted on flow-generated noise in ventilation systems. Because flow-generated noise can be severe at a high flow velocity, it has gained the interest of many scientific and technical investigators. Their experimental work focusing on data collection for flow-generated noise will first be reviewed, followed by those on the development of generalised predictive techniques for flow-generated noise.

### 3.3.1 Work on the collection of measured data for various duct elements

Various potential sources of flow-generated noise exist in ventilation ductwork systems. The following are considered.

Plain straight ducts, bends, take-offs, abrupt and gentle transformation pieces, tie rods, butterfly dampers, opposed blade dampers, damper-damper interaction, attenuators, terminal units, diffusers and associated dampers, grilles, high-speed jet outlets, and nozzles.

#### 3.3.1.1 Straight empty duct

Ingard et al. [19] carried out experiments in which air was passed through an empty 600 mm square mild steel duct at velocities of 10, 15, and 25 m/s.

Their results showed that the generated sound power level increased by approximately 18 dB per doubling of the flow speed. The energy in the spectrum decreased with the increase in frequency. However, an increase in the sound power was observed when the sound wavelength was comparable to the duct width.

Soroka [20, 21] investigated circular ducts with diameters between 100 and 250 mm, and rectangular/square ducts with equivalent cross-sectional areas, all open-ended. His results showed systematic relationships between the duct velocities and measured sound levels. Circular ducts were observed to be quieter than square ducts at lower velocities. Nevertheless, the rate of increase of the noise with velocity was higher for circular ducts. Thus, the circular ducts were nearly as noisy as the square ducts at the highest velocities tested. In addition, the rectangular duct was noisier than the square duct with the same cross-sectional area.

Soroka also measured the A-weighted sound levels in each case at approximately 600 mm (2 ft) from the centre of the exit section at an angle of 45°. Because only two rectangular ducts were tested, the data for the other duct sizes were obtained by extrapolation. Moreover, the analysis ignored the noise spectrum, which is important. Therefore, the uncertainty created by his approach and the abovementioned omissions have vitiated the authority of his work.

### 3.3.1.2 Butterfly dampers

Ingard et al. [19] measured the spectra of the noise generated by a 597 mm × 597 mm × 3 mm steel damper installed in a 600 mm square duct at an angle of 0°, 15°, and 45° with respect to the flow axis. The flow speeds ranged from 4 to 30 m/s in a series of experiments. For small-angle settings, the spectrum shape was similar to that of the empty duct spectra. However, the spectrum was remarkably flat at the highest angle setting. This change was attributed to the increase in the size and velocity of the turbulent wake that formed behind the plate as the angle setting of the damper increased.

### 3.3.1.3 Cylindrical rod in a straight duct

It is common to have a strut or sensor inside a ventilation duct system. Ingard et al. [19] tested a series of small-diameter rods in a 600 mm × 600 mm steel duct. The rods typically did not radiate any noticeable noise above the general background noise from the duct. Occasionally, however, the sound intensity was considerably amplified when the eddy frequency of the Karman vortices shed by the cylinder (Figure 3.4) coincided with the frequency of the lateral acoustic wave mode in the duct. Their results showed the effects of placing a 13 mm rod in a duct flow of 25 m/s. The increased noise from regeneration was preferentially

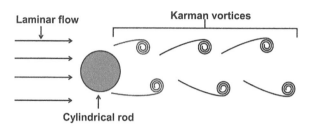

Laminar flow

Karman vortices

Cylindrical rod

*Figure 3.4* A schematic of the generation of Karman vortices due to a cylindrical rod.

amplified by coupling to a resonant mode of the 600 mm square duct at 500 Hz of noise from a 600 mm × 600 mm duct with and without four 13 mm diameter rod.

### 3.3.1.4 Duct transition sections

Ingard et al. [19] measured the sound power generated by three transition combinations using abrupt and gradual transformations. They found that a higher airflow rate produced a higher sound level. Nevertheless, the sound levels produced differed throughout the spectrum on a case-to-case basis. Basically, the abrupt transition pieces generated more sound than the gradual ones, but the change in the sound level produced had no direct relationship with the change in the cross-sectional area. Furthermore, with an area transition ratio of 3:1, an improvement of only a few decibels was achieved with the use of a gradual transition. For the smaller area ratio change of 1.5:1, an improvement of approximately 20 dB was achieved. With a large area change of 9:1, the results showed improvements approaching 30 dB with the use of gradual transition.

### 3.3.1.5 Bends

Kerka [22] was among the earliest investigators to measure the attenuation and noise generation by bends, both with and without turning vanes. He found that vaned bends were significantly noisier than unvaned bends. For vaned bends of the same size but different angles, the noise levels in the low- and mid-frequency range were similar for a given air velocity. The noise level at low frequencies for vaned bends of the same angle significantly increased as their size increased. For vaned bends of different aspect ratios (with opposite sides remaining constant in size), the noise level was nearly the same in the mid-frequency range. However, it increased in the low-frequency rage, and to a lesser extent in the high-frequency range, as the aspect ratio increased.

Ingard et al. [19] also measured the noise generated by bends. Six bends were tested during the experiments. Curved and rectangular 90° bends were tested with both square and rectangular cross-sections. The rectangular bends were tested with and without circular arc-turning vanes. Tests were conducted for each bend at various duct air velocities, reaching approximately 30 m/s with 400 mm square bends and 13 m/s with 600 mm × 200 mm rectangular bends. They found that the radiused bends generated marginal sound for the entire range of air flow rates considered. Therefore, this type of bend can generally be ignored as a source of significant aerodynamic sound.

Soroka [20, 21] conducted a series of experiments to test 90° sharp and radiused bends with and without terminating ducts, fitted to ducts with diameters ranging between 100 and 250 mm. He measured the A-weighted sound levels at a distance of 600 mm from the centre of the exit and at an angle of 45°. In some cases, additional measurements were performed at a distance of 900 mm for comparison. His objective was to measure the sound power in the direct field of the exit plane. Soroka performed his analysis solely on the basis of the total sound power generated; no reference was made to the noise spectrum. This approach is misleading, and the results lack vital information. Therefore, his conclusions should be interpreted with caution.

Watson [23] tested 725 mm × 350 mm and 350 mm × 750 mm lined and unlined ducts and bends, with and without airflow. His objective was to investigate the attenuation (insertion loss), self-generated noise, and pressure drop of different duct elements. Mitre bends without vanes, with standard guide vanes, and with acoustically treated guide vanes were tested. Watson's work had a limited scope and was not a systematic investigation. However, he observed that when he used specially lined double-skin construction guide vanes, the high-frequency attenuation effect of the vanes was significantly modified, and the performance appeared to be close to that of bends without vanes. The self-generated noise of the specially treated bends appeared comparable with that of any well-designed vaned mitre bend. The sound power levels obtained from the tests agreed closely with those obtained by Kerka [22] on similar elements.

### 3.3.1.6 Branch take-off sections

Brockmeyer [24] conducted an experiment to measure the sound power leaving the end of a 100 mm branch duct for two main flow velocities (9 and 18 m/s) and two junction designs (sharp and coned). The experimental results considered the duct 'end reflection' and indicated a substantial fall-off below 250 Hz. The term 'end reflection' is a phenomenon that occurs when the sound energy reaches the end of a duct run. A considerable amount of low-frequency energy is reflected upwards through the ductwork. If the outlet is small, then the amount reflected could reach 90% or more. This occurs because, for low-frequency sound, whose wavelength is significantly greater than that of the duct, the noise or pressure fluctuation compresses a slug of air in the pipe like a wave. Upon reaching the end of the pipe, the slug of air can suddenly expand sideways, and the resistance to the pressure-pulse motion disappears, which causes a considerable amount of air to exit the end of the pipe, and a negative pulse to travel back up the pipe, taking some of the energy with it.

### 3.3.2 Work on generalised predictive technique for flow-generated noise

The data obtained from the previous experiments described above can only be effectively applied to systems with the same configurations. Moreover, only general conclusions can be drawn. As an alternative to the establishment of a database of flow-generated sound powers, because of the variety of elements, attempts have been made to develop a more generalised prediction method.

Chaddock [25] experimentally found that the total sound power level in decibels $W_D$ of circular ceiling diffusers could be expressed by the following equation:

$$W_D = 32 + 13\log_{10} A_m + 60\log_{10} V_m \tag{3.4}$$

where
$A_m$ = minimum flow area in the neck, m$^2$ and
$V_m$ = velocity at the minimum area, m/s.

Hubert [26] plotted the total noise output of various diffusers against the resistance based on the concept of diffuser resistance, which he previously derived [27]. The empirical relationship he obtained is as follows:

$$W_D = L_s + 10\log_{10} A_m + 60\log_{10} V_m + 30\log_{10} \xi, \tag{3.5}$$

where
$W_D$ = sound power level, dB;
$L_s$ = specific sound power level (re $10^{-12}$W);
$\xi$ = specific resistance of the diffuser;
$A_m$ = minimum flow area in the neck, m$^2$; and
$V_m$ = velocity at the minimum flow area, m/s.

The concept of a *pressure-based* scheme to predict the sound generated by the interaction between a flow and surface was first proposed by Iudin [28]. He began with the assumption that the acoustic power of flow-generated noise, and its distribution over the frequency spectrum, are determined by the geometric shape of the duct, its dimensions, flow parameters, and the acoustic properties of the volume in which the sound was generated. He further assumed that the flow parameters in an air duct with a specific geometric shape are determined by the static pressure differential $P$ across it; the physical properties of the medium (density $\rho$ and sound velocity $c$); and dimensionless constants (Reynolds number $Re$ and Mach number $M$).

$$Re = \frac{Vd_c}{\nu} \tag{3.6}$$

$$M = \frac{V}{c} \tag{3.7}$$

where
$V$ = time-averaged flow velocity at a characteristic point in the air duct;
$d_c$ = determined geometric dimension (representative dimension); and
$\nu$ = kinematic viscosity of the medium.

According to Iudin, the acoustic properties of the volume are determined by the dimensionless impedances $Z$ of the boundaries, and by the relationship between the sound wavelength and dimensions of the volume $\lambda/d_c$. The frequency of the aerodynamic sound is determined by the Strouhal number $St$.

$$St = \frac{fd}{V} = Q_1 \left( \text{the geometric shape}, Re, M \right) \tag{3.8}$$

Because the frequency is $f = c/\lambda$, then $\lambda/d_c = 1/(MSt)$; therefore, $\lambda/d_c$, which is a constant when $Z$ is also a constant, is not a determining criterion.

Based on these assumptions, the acoustic power of the air duct can be expressed as follows:

$$W = Q_2 \left( \Delta p_s, \rho, c, d_c, Re, M, Z, \frac{\lambda}{d} \right) \tag{3.9}$$

Using the dimensional analysis of the source mechanism, he obtained a simple formula from his assumptions:

$$\frac{W}{\rho c^3 d^2} \left( \frac{\Delta p_s}{\rho c^2} \right)^{\alpha} = Q_3 \left( \text{shape}, Re, M, Z \right) \tag{3.10}$$

This equation is the working similarity formula for his experiments.

Iudin studied the noise of various metal air duct elements, including low- and high-pressure ejectors, powdery granular material layers, and aerodynamic velocity pipes. The investigations were performed in both acoustic and untreated rooms and in open air. By combining the results of his studies with a dimensional analysis of the source mechanism, he concluded that the acoustic power is proportional to the cube of the excess pressure and the square of the geometric dimension.

Iudin's work has been relatively neglected in favour of the work of Gordon [29, 30]. Initially, Gordon and Maidanik [31] studied the sound-generating capabilities of a flow spoiler in a pipe environment using a scale model. Their investigation of the effect of upstream flow discontinuities on the acoustic power radiated by an air jet was expanded by Gordon [29, 30]. The analysis by Gordon was based on the assumption that the magnitude of the fluctuating forces associated with an aerodynamic source is proportional to the steady-state drag forces. He investigated the sound power generated by various spoilers close to the end of a pipe carrying a high-velocity air flow. He [29] initially derived the following empirical formula to predict the total acoustic power generated by a spoiler in a duct.

$$ W = \frac{K \left( p_0 - p_a \right)^3 D^2}{\rho_a^2 c_a^3}, \tag{3.11} $$

where
$W$ = acoustic power;
$p_0$ = total stagnation pressure on the upstream side of the spoiler (measured with a pitot static tube);
$p_a$ = atmospheric pressure;
$\rho_a$ = atmospheric density;
$c_a$ = velocity of sound;
$D$ = duct diameter; and
$K$ = constant with an experimentally determined value of $2.5 \times 10^{-4}$.

The geometry of the spoiler was not directly input into Eq. (3.11) but was implicit in the pressure drop $(p_0 - p_a)$ across the flow spoiler.

The frequency spectrum of the flow-generated noise is crucial in engineering design. Gordon found that a normalised spectrum could be obtained by collapsing the data from a series of experiments onto different spoiler configurations. He then modified Eq. (3.11) into the following form:

$$ W = \frac{K \left( p_0 - p_a \right)^3 D^2}{\rho_a^2 c_a^3} \left[ 1 + \left( \frac{f_c}{f_0} \right)^2 \right], \tag{3.12} $$

where
$f_0$ = constant frequency related to the dimension of the duct spoiler system;
$f_c$ = octave band centre frequency; and
$D$ = duct diameter.

He observed that $f_0$ tended to be close to the cross-mode onset frequency, and collapsed his data based on this correction.

Gordon considered air velocities and pressure drops across spoilers in excess of those encountered in ventilation systems. Although his work cannot be applied to predict flow-generated noise in ventilation systems, his work along with that of Iudin, support the idea that a pressure-based prediction technique may be applicable to conditions encountered in ventilation systems. Although he devised scaling laws to collapse his measured sound power data onto a 'generalised spectrum', he derived his scaling laws from a free-field radiation model, in which the noise was generated by a point fluctuating source and the effect of the duct was neglected.

In 1970, Heller and Widnall [32] presented the results of their theoretical and experimental study on the correlation between fluctuating forces on rigid flow spoilers with the corresponding sound radiation. As an extension of the experimental work of Gordon and Maidanik [31], they directly measured the fluctuating drag and lift forces on flow spoilers. They then demonstrated a direct correlation between the fluctuating forces and radiated sound under both free-field and confined-environment conditions. Their data correlated well when they plotted $[20log_{10}(F_{drag}/F_0) - 40log_{10}(U/U_0)]$ and $[20log_{10}(P/P_0) - 60log_{10}(U/U_0)]$ against the Strouhal number $St = f_c d/U$ [32, 33]. $F_{drag}$ is the measured fluctuating drag force (N), $F_0$ is the reference force ($1 \times 10^{-5}$ N), $U$ is either the jet exhaust velocity or in-pipe flow velocity (m/s), $U_0$ is the reference velocity (1 m/s), $f_c$ is the 1/3 octave band centre frequency, $d$ is the length dimension related to a typical dimension of the spoiler, $P$ is the sound pressure, and $P_0$ is the reference pressure.

A theory, which considers the effects of the enclosure on the sources and the pipe-end reflection, was developed to predict the sound power radiated to the free field from pipe-immersed flow spoilers. In this theory, the efficiency of the sound power radiation from dipole sources in a hard-walled pipe (confined environment) increased by a frequency-squared term, which changed the sound power/flow velocity dependence to a quadrupole dependence, i.e., to an eighth-power-law dependence. However, the effect of end reflection introduced an inverse frequency-squared term that restored the original sixth power of the velocity-dependence of the dipole-source power radiation. In addition, the sound power radiated by dipole sources in a confined environment increased by a factor of three, as predicted and observed in their experiments. Hence, they clarified the acoustical signifi-cance of the duct enclosing the noise source.

In 1973, Holmes [34, 35] extended Gordon's work to develop a tech-nique that became the basis of a flow-generated noise prediction method owing to the variety of in-duct elements. He investigated the noise generated by supply grilles, dampers, grille and damper combinations, and bends fitted with turning vanes. He considered the effect of the ductwork enclosing an airstream. He assumed that the acoustic power generated was proportional to the product of the area and the sixth power of the air velocity. The general technique obtained the effective blockage factor $B_e$ (ratio of the free area and duct area) from the pressure loss data $C_L$. This value was then used to calculate the free area and velocity. Finally, the general spectra were employed to obtain the frequency spectrum for a given condition.

The relationships proposed for the sound produced by air duct elements are as follows.

Damper:

$$L_w = K_1 + K_V + K_B + K_s, \, dB \, ref \, 10^{-12} \, W \qquad (3.13)$$

where
$K_1$ = general damper spectrum;
$K_V$ = velocity correction factor;
$K_B$ = duct correction factor; and
$K_S$ = spectrum correction factor.

Similar calculations were performed for grilles and bends with turning vanes using the appropriate charts and tables.

This prediction method was developed from a simplified theory and a limited range of in-duct components and duct sizes; hence, it may not be applicable to systems with different configurations.

Nelson and Morfey [36] investigated the aerodynamic sound production in low-speed flow ducts. In developing their theory, they considered the effect of the duct environment on noise generation. The basis of their theory is that the sound power radiated by an in-duct spoiler is related to the total fluctuating drag force acting on the spoiler, which is a function of the turbulence intensity in the spoiler region. Because they were unable to determine the actual spectrum of the turbulence intensity within the vicinity of the spoiler, to arrive at a predictive technique, they further assumed that the fluctuating drag force is directly proportional to the steady drag force. (The same assumption was made by Gordon [30] in devising his theory, and its validity was confirmed by the experiments of Heller and Widnall [32].) The collapse of the experimental data onto a generalised spectrum to form the basis of the predictive technique was achieved by the empirical evaluation of the constant of proportionality between the fluctuating and steady drag forces as a function of the Strouhal number.

The Nelson–Morfey equations for determining the sound power generated by an in-duct spoiler are as follows.

For the centre frequency below the cut-on frequency of the first transverse duct mode $f_c < f_o$

$$120 + 20 \log_{10} K(St)$$

$$= L_{\text{w,D}} - 10 \log_{10} \left[ \frac{\rho_0 A \left( \left\{ \sigma^2 (1-\sigma) \right\} \right)^2 C_\text{D} U_\text{c}^4}{16 c_0} \right] \qquad (3.14)$$

For the centre frequency above the cut-on frequency of the first transverse duct mode $f_\text{c} > f_\text{o}$

$$120 + 20 \log_{10} K(St)$$

$$= L_{\text{w, D}} - 10 \log_{10} \left\{ \frac{\rho_0 \pi A^2 \, (st)^2 \, [\sigma^2 (1-\sigma)^2 \, C_\text{D}^2 U_\text{c}^6}{24 c_0^3 d_\text{c}^2} \right\}$$
$$- 10 \log_{10} \left\{ 1 + \left( \frac{3 \pi c_0}{4 \omega_\text{c}} \right) \frac{(a+b)}{A} \right\} \qquad (3.15)$$

where
$L_{\text{w,D}}$ = in-duct sound power level;
$K(St)$ = single Strouhal number-dependent constant;
$\rho_0$ = density of air;
$A$ = cross-section area of the duct;
$\sigma = A_\text{c}/A$ = duct unobstructed area/duct area = open area ratio;
$St$ = Strouhal number = $f_\text{c} d_\text{c}/U_\text{c}$;
$U_\text{c}$ = $q/A_\text{c}$ = volume flow rate/duct unobstructed area = velocity in the restriction;
$c_0$ = ambient speed of sound;
$d_\text{c}$ = characteristic dimension of the spoiler;
$\omega_\text{c}$ = angular centre frequency of the band of frequencies under consideration;
$a$ = duct width;
$b$ = duct height;
$C_\text{D}$ = drag coefficient = $F_3/(\frac{1}{2}\rho_0 \, U^2 A (1-\sigma))$; and
where $F_3$ = steady-state force on the spoiler.

All the terms in the proposed predictive equations are constants or meas-
urable variables, plus a single Strouhal number-dependent constant. When
the value of the Strouhal number-dependent constant is established, it is
possible, in principle, to employ the Nelson–Morfey theory for predictive
purposes.

The experimental results presented by Nelson and Morfey were obtained
using simple flat-plate flow spoilers in square-section ductwork. These
cannot easily be related to typical elements found in real ventilation systems.
With this type of spoiler, the value of the characteristic dimension can
be easily determined. To apply the Nelson–Morfey theory to other duct

obstructions, the appropriate values of this parameter must be determined. For obstructions other than flat-plate spoilers in a square duct or for any obstruction in a circular duct, the values of this parameter cannot be determined by simply inspecting the geometry of the situation.

Oldham and Ukpoho [37] extended the work of Nelson and Morfey to the case of circular ductwork and more complex flow spoilers. They rewrote the Nelson–Morfey equations by determining the appropriate values of the open area ratio and the characteristic dimension to apply the work of Nelson and Morfey to more complex flow spoilers in circular or square ducts.

The modified Nelson–Morfey equations by Oldham and Ukpoho for determining the sound power generated by an in-duct spoiler are as follows:

For the centre frequency $f_c < f_o$

$$120 + 20\log_{10} K(St)$$

$$= L_{w,D} - 10 \log_{10} \left[ \frac{\rho_0 A \sigma^4 C_L^2 U_c^4}{16 c_0} \right] \tag{3.16}$$

For the centre frequency $f_c > f_o$

$$120 + 20\log_{10} K(St)$$

$$= L_{w,D} - 10 \log_{10} \left\{ \frac{\rho_0 \pi A^2 (st)^2 \sigma^4 C_L^2 U_c^6}{24 c_0^3 d_c^2} \right\} - 10 \log_{10} \left[ 1 + \frac{3c_0}{8 r f_c} \right] \tag{3.17}$$

where

$$d_c = \frac{\pi r (1 - \sigma)}{2} \tag{3.18}$$

and

$$St = \frac{f_c \pi r (1 - \sigma)}{2 U_c} \tag{3.19}$$

where
$L_{w,D}$ = in-duct sound power level;
$K(St)$ = single Strouhal number-dependent constant;
$\rho_0$ = density of air;
$A$ = cross-section area of the duct;
$A_c$ = duct unobstructed area;
$\sigma = A_c/A$ = duct unobstructed area/duct area = open area ratio;
$C_L$ = pressure loss coefficient;

$U_c = q/A_c$ = volume flow rate/duct unobstructed area = velocity in the restriction = $U/\sigma$;
$c_o$ = ambient speed of sound;
$d_c$ = characteristic dimension of the spoiler;
$f_c$ = centre frequency of the band of frequencies under consideration;
$St$ = Strouhal number = $f_c d/U_c$; and
r = radius of the circular duct.

Oldham and Ukpoho conducted experiments on dampers and orifice plates in a circular duct to produce a generalised spectrum by collapsing the experimental data on the basis of the modified Nelson–Morfey equations. The spectra obtained using dampers and orifice plates as spoilers were similar to those obtained by Nelson and Morfey with simple strip spoilers. Although the results of Oldham and Ukpoho lend further support to the concept of a generalised prediction method based on pressure loss, their experiments were carried out using ductwork of a similar size to that of Nelson and Morfey. Thus, the generalised prediction curve they derived may not apply to systems with very different duct dimensions.

Oldham and Ukpoho [38] also investigated the interactions between closely spaced duct elements. Their experimental data showed that placing a second element in the downstream turbulence of another element increased the noise levels experienced with both elements in relative isolation. They also found that the increase in noise level was frequency-dependent and a function of the spoiler separation and duct diameter. The former effect resulted from the acoustic reflection between the two elements, which is a function of their acoustic impedances. Despite their studies on the interaction between two closely spaced duct elements, their modified Nelson–Morfey predictive equations are similar to those from previous techniques, and are only applicable to an isolated element in air ducts.

Most predictive methods, including those in the *CIBSE Guide* [2] are only applied to an isolated in-duct element, which is very different from that found in practical systems. Mak and Yang [39, 40] developed a predictive technique for the aerodynamic noise radiated by two elements in air ducts. To develop their theory, they used a simple source model to represent the partially coherent noise sources. Then, they also hypothesised a constant of proportionality between the fluctuating and steady-state drag forces acting on the flow spoilers. Based on the work of Nelson and Morfey, Mak and Yang finally developed a predictive technique for the flow-generated noise produced by two in-duct flow spoilers. Thus, the inferred infinite-duct values of the radiated sound power level $L_{w,D}$ in the frequency band can be normalised by evaluating the following:

For the centre frequency below the cut-on frequency of the first transverse duct mode $f_c < f_0$

$$120 + 20\log_{10} K(St)$$

$$= L_{w,D} - 10\,\log_{10} \left\{ \frac{\rho_0 A^2 \left[ \sigma^2 (1-\sigma) \right]^2 C_D^2 U_c^4}{16 c_0} \right\}$$

$$- 10\,\log_{10} \left\{ 1 + 2\sqrt{\gamma_{12}^2} \cos\left( \frac{\omega_c d}{c_0} \right) \cdot \cos\left[ \phi(\omega_c) \right] \cdot \xi + \xi^2 \right\}$$

(3.20)

For the centre frequency below the cut-on frequency of the first transverse duct mode $f_c > f_0$

$$120 + 20\log_{10} K(St)$$

$$= L_{w,D} - 10\,\log_{10} \left\{ \frac{\rho_0 \pi A^2 (St)^2 \left[ \sigma^2 (1-\sigma) \right]^2 C_D^2 U_c^6}{24 C_0^3 d_c^2} \right\} \times \left[ 1 + \left( \frac{3\pi c_0}{4\omega_c} \right) \frac{(a+b)}{A} \right]$$

$$- 10\,\log_{10} \left\{ 1 + 2\sqrt{\gamma_{12}^2} \cos\left( \frac{\omega_c d}{c_0} \right) \cdot Q \cdot \cos\left[ \phi(\omega_c) \right] \cdot \xi + \xi^2 \right\}$$

(3.21)

where
$a, b$ = duct cross-section dimensions;
$A$ = duct cross-sectional area;
$c_0$ = ambient speed of sound;
$C_D$ = drag coefficient;
$d$ = distance between two flat plate spoilers;
$f_0$ = cut-on frequency of the first transverse duct mode;
$f_c$ = centre frequency of the measurement band;
$K(St)$ = constant of proportionality between the root-mean-square fluctuating drag force and steady-state drag force;
$Q$ = a constant;
$d_c$ = characteristic dimension, width of the rectangular spoiler;
$St$ = Strouhal number;
$L_w$ = in-duct sound power level due to two elements;
$U_c$ = flow velocity in the duct constriction;
$\rho_0$ = ambient air density;
$\sigma$ = open area ratio;
$\omega_c$ = centre angular frequency of the measurement band;
$\zeta$ = ratio of mean drag forces acting on the spoilers;
$\phi(\omega_c)$ Phase angle of cross-power spectral density of source volumes; and
$\gamma_{12}^2$ = coherence coefficient.

These predictive equations were later revised [41] to compare their results with the experimental results of Oldham and Ukpoho. The predicted results generally agreed with the experimental results of Oldham and Ukpoho.

However, the technique proposed by Mak and Yang is limited to two in-duct elements. Because there are always multiple in-duct elements (more than two elements) in a practical ventilation ductwork system, the technique was further developed theoretically by Mak [42] and Mak et al. [43, 44] to predict the flow-generated noise produced by multiple in-duct elements. These methods consider the acoustical and/or aerodynamic interaction(s) of multiple flow-noise sources in an air duct. Mak et al. [45] experimentally confirmed the effectiveness of the prediction method of Mak [42] for multiple in-duct spoilers. Two predictive equations [42, 45] were obtained to determine the radiated sound power generated by the interaction between multiple spoilers, one corresponding to the centre; frequencies$f_c$ below the cut-on frequency$f_0$, and another to the centre frequencies above it. For $N(N > 2)$ elements:

For $f_c < f_0$

$$\Pi_N = K^2(St) \times \Gamma_1 \times I_1 \tag{3.22}$$

For $f_c > f_0$

$$\Pi_N = K^2(St) \times \Gamma_2 \times I_2, \tag{3.23}$$

where
$\Pi_N$ = infinite-duct values of the radiated sound power owing to multiple $(N)$ spoilers;
$K(St)$ = ratio of fluctuating and steady-state drag forces on the spoilers;
$K^2(St)$ = square of the ratio $K(St)$;
$St$ = Strouhal number determined by $St = f_c d_c/U_c$;
$d_c$ = characteristic dimension of the in-duct element (m);
$f_c$ = centre frequency of measurement band (Hz); and
$U_c$ = flow velocity in the constriction (m/s).

The power term $\Gamma_1$ from the first spoiler below the cut-on frequency is:

$$\Gamma_1 = \left\{ \rho_0 A \left[ \sigma^2 (1 - \sigma) \right]^2 C_D^2 U_c^4 / 16 c_0 \right\}$$

The power term $\Gamma_2$ from the first spoiler above the cut-on frequency is:

$$\Gamma_2 = \left\{ \rho_0 \pi A^2 (S)^2 \left[ \sigma^2 (1-\sigma) \right]^2 \times \left[ C_D^2 U_c^6 / 24 c_0^3 r^2 \right] \right.$$
$$\left. \times \left[ 1 + (3\pi c_0 / 4\omega_c)(a+b)/A \right] \right\}$$

The interaction term $I_1$ below the cut-on frequency is:

$$I_1 = \left\{ \sum_{i=1}^{N} \zeta_i^2 + 2 \sum_{i=1}^{N-1} \sum_{j=1}^{N-2} \left[ \sqrt{\gamma_{i(i+1)}^2} \times \left[ \cos(\omega_c d_{i(i+1)} / c_0) \cos[\phi_{i(i+1)}(\omega_c)] \zeta_i \zeta_{i+1} \right. \right. \right.$$
$$\left. \left. \left. + \sqrt{\gamma_{(i-1)(i+j)}^2} \times \cos(\omega_c d_{(i-1)(i+j)} / c_0) \cos[\phi_{(i-1)(i+j)}(\omega_c)] \zeta_{i-1} \zeta_{i+j} \right] \right\} \right.$$

The interaction term $I_2$ above the cut-on frequency is:

$$I_2 = \left\{ \sum_{i=1}^{N} \zeta_i^2 + 2 \sum_{i=1}^{N-1} \sum_{j=1}^{N-2} \left[ \sqrt{\gamma_{i(i+1)}^2} \; Q_{i(i+1)} \times \cos[\phi_{i(i+1)}(\omega_c)] \zeta_i \zeta_{i+1} \right. \right.$$
$$\left. \left. + \sqrt{\gamma_{(i-1)(i+j)}^2} Q_{(i-1)(i+j)} \cos[\phi_{(i-1)(i+j)}(\omega_c)] \zeta_{i-1} \zeta_{i+j} \right] \right\}$$

where $N > 2$; $N$, $i$, and $j$ are integers and $j = 1, 2, ..., (N-2)$.
   In the above equation, a value is ignored if its subscript is zero or greater than $N$.

$Q_{i(i+1)}$ is given by $Q_{i(i+1)} = \Omega_{i(i+1)} \Big/ \Psi_{i(i+1)}$,

where

$$\Omega_{i(i+1)} = \frac{k^2 ab}{6\pi} 3 \left[ \frac{\sin e}{e} + \frac{2 \cos e}{e^2} - \frac{2 \sin e}{e^3} \right] + \frac{k(a+b)}{8} 2 \left[ J_0(e) - \frac{J_1(e)}{e} \right],$$

$$\Psi_{i(i+1)} = \left[ \frac{k^2 ab}{6\pi} + \frac{k(a+b)}{8} \right],$$

and $e = k d_{i(i+1)}$;

$Q_{(i-1)(i+j)}$ is given by $Q_{(i-1)(i+j)} = \Omega_{(i-1)(i+j)} \Big/ \Psi_{(i-1)(i+j)}$,

where

$$\Omega_{(i-1)(i+j)} = \frac{k^2 ab}{6\pi} 3\left[\frac{\sin e}{e} + \frac{2\cos e}{e^2} - \frac{2\sin e}{e^3}\right] + \frac{k(a+b)}{8} 2\left[J_0(e) - \frac{J_1(e)}{e}\right],$$

$$\Psi_{(i-1)(i+j)} = \left[\frac{k^2 ab}{6\pi} + \frac{k(a+b)}{8}\right],$$

and $e = kd_{(i-1)(i+j)}$;

$d_{ij}$ is the distance between the $i$th and $j$th spoiler; $k$ is the wave number; $J_0$ and $J_1$ are the zero- and first-order Bessel's functions, respectively; $a$ and $b$ are the duct cross-section dimensions; $d_c$ is a characteristic dimension of the element; $A$ is the area of the duct cross-section by $A = (a \times b)$; and $A_c$ is the area of the duct constriction by $A_c = (a \times b - d_c \times b)$. $U_c$ is the flow velocity in the constriction provided by the spoiler, and is defined by the volume flow rate $q$ and the area of the duct constriction $A_c$, such that $U_c = \dfrac{q}{A_c} = \dfrac{UA}{A_c}$.

$U$ is the mean duct flow velocity, and $\sigma$ is the open area ratio determined by $\sigma = A_c/A$. The Strouhal number for this duct and flow is $St = f_c d_c/U_c$; the factor $K(St)$ is the ratio of fluctuating to steady-state drag forces on the spoilers; $P_N$ is the infinite-duct values of the radiated sound power; $f_0$ is the cut-on frequency of the first transverse duct mode (i.e., the least non-zero value of the cut-on frequency is defined by $f_0 = (c_0/2\pi)\sqrt{(m\pi/a)^2 + (n\pi/b)^2}$, where $m,n = 0,1,\ldots$); $c_0$ is the ambient speed of sound; $\rho_0$ is the ambient air density; $\gamma^2_{ij}$ is the coherence function of the $i$th and $j$th spoilers; $\omega_c$ is the centre radiant frequency of the measurement band; $\phi_{ij}(\omega_c)$ is the phase of the cross-power spectral density of the source volume of the $i$th and $j$th sound sources; $\xi_i$ is a constant ratio of the mean drag forces acting on the $i$th and the first spoilers; $\Delta P_s \sqrt{2}$ is the static pressure drop across a spoiler (Pa); and $C_D$ is the drag coefficient determined by

$$C_D = \frac{\Delta P_s}{\frac{1}{2}\rho_0 U_c^2 \sigma^2 (1-\sigma)} \tag{3.24}$$

By comparing the above expressions with those obtained by Nelson and Morfey [36] for the sound power generated by an isolated in-duct spoiler, the interaction factor $\beta_N$ can be defined as follows:

$$\beta_N = \begin{cases} I_1, & f_c < f_0 \\ I_2, & f_c > f_0 \end{cases} \tag{3.25}$$

Furthermore, if the sound power due to an in-duct spoiler is denoted as $\Pi_s$, then a simple relationship between $\Pi_N$, the sound power due to multiple ($N$) spoilers, and that due to a single spoiler is obtained as follows:

$$\Pi_N = \Pi_s \times \beta_N \tag{3.26}$$

where $\Pi_s$ can be obtained using the prediction method proposed by Nelson and Morfey [36], and $\beta_N$ can be determined experimentally.

$\xi_i$ is the constant ratio of the mean drag forces acting on the $i$th and the first spoilers. The first spoiler is closest to the inlet of the air flow:

$$\xi_i = \frac{\overline{F_{z1}}}{\overline{F_{zi}}} \tag{3.27}$$

where $\overline{F_{zi}}$ is the mean drag force acting on the $i$th spoiler counted from the inlet of the air flow, and $\overline{F_{z1}}$ is the mean drag force acting on the first spoiler. The mean drag force acting on the $i$th spoiler can be expressed as $F_{zi} = A\Delta P_s$.

The phase of the cross-power spectral density of the source volume of the $i$th and $j$th sound source $\phi_{ij}\left(\omega_c\right)$ can be given by

$$\phi_{ij}\left(\omega_c\right) = \delta_{ij} - kM\overline{d_{ij}}, \tag{3.28}$$

where $\delta_{ij}$ is the difference between the phases of the total fluctuating drag forces acting on the $i$th and $j$th spoiler: $\delta_{ij} = \theta_j\left(\omega\right) - \theta_i\left(\omega\right)$, where $\theta_i\left(\omega\right)$ and $\theta_j\left(\omega\right)$ are the phases of the fluctuating drag force acting on the $i$th and the $j$th spoiler, respectively. $k$ is the wave number, $M = U/c_0$, and $\overline{d_{ij}} = d_{ij}/\left(1 - M^2\right)$.

The coherence functions $\gamma^2_{ij}$ of the noise sources is given by

$$\gamma^2_{ij} = \log^2_{10}\left(1 + \frac{15}{Ud_{ij}} \frac{\left|F_j(\omega)\right|^2}{\left|F_i(\omega)\right|} \cos^2\left(\theta_j(\omega) - \theta_i(\omega)\right)\right) \tag{3.29}$$

where $F_i\left(\omega\right)$ and $\theta_i\left(\omega\right)$ are the magnitude and phases of the fluctuating drag force acting on the $i$th spoiler, respectively.

The main drawback of such a prediction method for multiple in-duct elements is that it requires the ratio of the mean drag forces, phase relationship between the fluctuating drag forces acting on the spoilers, and coherence function of the noise sources. Determining these parameters requires expensive acoustic and aerodynamic facilities.

In this chapter, we mentioned that there is only a limited amount of measured data on the flow-generated noise produced by some in-duct elements, such as bends and transition pieces. This lack of data reflects

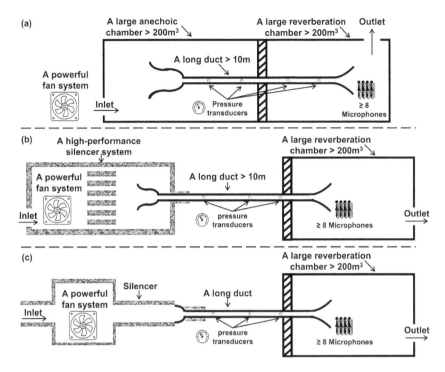

*Figure 3.5* Schematic diagrams of the experimental setup for flow-generated noise measurements using reverberation chamber methods in the studies of (a) Oldham and Ukpoho [37]; (b) Nelson and Morfey [36]; and (c) Mak et al. [45].

the difficulty in obtaining such information by conventional measurement techniques that require the use of an expensive and specially combined acoustic and aerodynamic experimental facility consisting of the following:

1.  a powerful fan system;
2.  a high-performance silencer system to reduce the noise from the fan entering the test section of the ductwork;
3.  a long test duct to ensure stable flow conditions at the test element;
4.  a large reverberant room with a low level of background noise in which the sound power generated can be measured. Examples include the experimental setup of Nelson and Morfey [36], Oldham and Ukpoho [37], and Mak et al. [44] (Figure 3.5). Other researchers have used a large anechoic chamber for acoustic measurements [5, 29].

An alternative to using specialised and aerodynamic facilities currently gaining support in building engineering is using *computational fluid dynamics* (CFD) software packages, which is a powerful design tool that can predict the behaviour of fluid flow regimes. It has been applied to several areas of building engineering with considerable success, such as smoke extraction. The use of the pressure drop across in-duct elements to predict the sound power level of flow-generated noise is attributed to the fact that investigators have been unable to determine the actual spectrum of the turbulence intensity within the vicinity of a spoiler. However, flow-generated noise originates from the turbulence itself and its interaction with an acoustically hard surface. The use of CFD enables the direct determination of the turbulent field within the vicinity of the spoiler. Therefore, the validity of the basic assumption regarding the constant of proportionality between fluctuating and steady drag forces, which is the basis of pressure-based techniques, can be investigated. A representative dimension for dissimilar in-duct elements can also be obtained by studying the turbulent field within the vicinity of the in-duct elements. In addition, a relationship between turbulence and acoustic energy can be established to develop a generalised predictive technique for flow-generated noise. CFD was employed by Mak and Oldham [46, 47] to develop their technique for predicting flow-generated noise in ventilation systems. This technique is based on the relationship between the turbulent kinetic energy and acoustic energy. Although they have developed a turbulence-based prediction technique for flow-generated noise, their technique, similar to that of other researchers, can only be used for an isolated element in an air duct. In a real system, acoustic and aerodynamic interactions always occur between nearby elements in an air duct.

To further develop the prediction methods of Mak et al. for practical applications, a CFD approach [48, 49] was adopted to determine the parameters of the prediction methods, such as the ratio of the mean drag forces, the phase relationship between the fluctuating drag forces acting on the spoilers, and the coherence function of the noise sources.

Bullock [50] transformed the test data obtained by Ingard et al. [19] into a more general and useful form. The results of his analytical work have been adopted by *ASHRAE Handbook* [1] for the estimation of flow-generated noise by an isolated air duct element.

The *CIBSE Guide B4 Noise and Vibration Control for HVAC* [2] proposed an equation (*CIBSE Guide B4 method*) to estimate the overall sound power level $L_w$ generated by an isolated duct fitting:

$$L_w = C + 10\log_{10} A + 60\log_{10} u \tag{3.30}$$

where
$C$ = a constant dependent on the fitting and flow turbulence;
$A$ = minimum flow area of the fitting, m²; and
$u$ = maximum flow velocity in the fitting, m/s.

Typical values of $C$ for a range of duct fittings in low-turbulence flow conditions and the corrections to obtain the octave band power levels are provided in *CIBSE Guide B4*.

In addition to the *CIBSE Guide B4* method, the *Guide* provides a generic formula based on the concept of the *pressure-based* scheme and the works of Nelson and Morfey, and Oldham and Ukpoho for predicting flow-generated noise produced by an isolated duct fitting. Two predictive equations for the flow-generated noise produced by a single isolated in-duct spoiler are used to obtain the sound power level $L_W$ in different frequency ranges (below and above the 'cut-on' frequency $f_0$ of the duct). The cut-on frequency $f_0$, which was determined by the geometries of the duct as $f_0 = c/2l$ for rectangular ductwork or $f_0 = 1.841c/2\pi r$ for circular ductwork (where $c$ is the velocity of sound, $l$ is the longest sectional dimension of the rectangular ductwork, and $r$ is the radius of the circular duct), determines the wave propagation modes through the duct. According to *CIBSE Guide B4*, the equations for the sound power level $L_W$ are given as [2]:

For $f_c < f_0$ (below the cut-on frequency)

$$L_w = -37 + 20\log_{10}\left(K(St)\right) + 20\log_{10}\xi + 10\log_{10}A + 40\log_{10}u \quad (3.31)$$

and for $f_c > f_0$ (above the cut-on frequency)

$$\begin{aligned} L_w = &-84 + 20\log_{10}\left(K(St)\right) + 20\log_{10}(St) + 10\log_{10}\xi \\ &- 40\log_{10}\sigma + 10\log_{10}A + 60\log_{10}u \end{aligned} \quad (3.32)$$

where $f_0$ represents the octave band centre frequency, $A$ is the cross-sectional area of the duct, $u$ is the air velocity in the duct, $\zeta$ is the pressure loss factor, $\sigma = \left(\xi^{1/2} - 1\right)/(\xi - 1)$ is the clear area ratio, and $K(St)$ is an experimentally determined factor related to the Strouhal number $St = af_c\sigma(1 - \sigma)/u$ ($a$ represents the duct height or duct diameter). Figure 3.6 shows the determination of the term $20\log_{10}(K(St))$ from the Strouhal number. The term $20log_{10}(K(St))$ can be obtained once the Strouhal number is derived. The pressure loss factor $\xi$ can be obtained from the expression $\xi = \Delta P/0.5\rho u^2$ (where $\Delta P$ is the static pressure drop due to the component, and $\rho$ is the density of air).

The flow-generated noise problem caused by in-duct elements is due to the complex acoustic and turbulent interactions between multiple in-duct flow noise sources. The prediction method of Mak et al. [42, 45] for multiple in-duct elements considered the effects of interactions between flow-generated noise sources in ventilation ductworks by introducing an interaction factor in the predictive equations in Eq. (3.25). This interaction factor, denoted as $\beta_m$, was later introduced [51] to revise the generic formula in *CIBSE Guide*

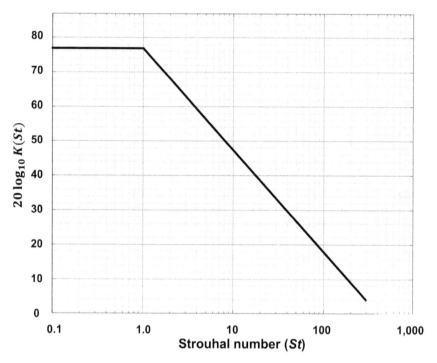

*Figure 3.6* Determination of term from the Strouhal number [2].

*B4* to develop a simplified and practical engineering prediction method for multiple in-duct elements. This simplified method is explained as follows.

The sound power generated by a single in-duct element can be obtained by Eq. (3.31) (for $f_c < f_0$) and Eq. (3.32) (for $f_c > f_0$) in different frequency ranges. For the sound power generated by multiple in-duct elements, a simple relationship between the sound power level owing to multiple elements and that owing to a single element is obtained as follows:

$$L_{WN} = L_w + 10\log_{10}\beta_N,$$  (3.33)

where $L_{WN}$ and $L_w$ represent the sound power level owing to multiple and a single element (obtained from Eq. (3.31) and Eq. (3.32)), respectively; and $\beta_{mN}$ is the interaction factor. Notably, $L_w$ is the element with the largest pressure loss factor $\xi$. Based on the earlier works of Mak et al. [42, 45], the simplified equations of the interaction factor are expressed as follows:

For $f_c < f_0$ (below the cut-on frequency)

$$\beta_N = N + N \cos KL \cos\left(kL - kL\frac{u}{c}\right) \tag{3.34}$$

and for $f_c > f_0$ (above the cut-on frequency)

$$\beta_N = \begin{cases} N + N \dfrac{8kA + \pi(a+b)}{2kA + \pi(a+b)}, & \text{for a rectangular duct} \\[2ex] N + N \dfrac{4kA + \pi r}{kA + \pi r}, & \text{for a circular duct} \end{cases} \tag{3.35}$$

where $N \geq 2$ is the number of in-duct elements, $k = 2\pi f_c/c$ is the wave number, $L$ is the shortest distance between two elements, $a$ and $b$ are the duct cross-sectional dimensions, and $r$ is the radius of the circular duct.

The interaction factor $\beta_N$ is primarily a function of the frequency, distance between the various duct elements, and geometries of the ventilation ductwork system. It can be observed from Eq. (3.33) that the $L_{WN}$ increased when the interaction factor $\beta_N > 1$, and decreased when $0 < \beta_N < 1$. However, Eq. (3.35) only considers the case of $\beta_N > 1$ by removing the product term $\cos(kL - kLu/c)$ from the second term on the right-hand side of Eq. (3.35). The purpose of such simplification is to prevent underestimations at high frequencies. The values of the interaction factor could be obtained in a straightforward manner at the design stage once the geometries of the ventilation ductwork system have been determined. Although the proposed interaction factor $\beta_N$ only represents the approximate added sound pressure level, and hence could overestimate it, the simple and convenient prediction formulae meet the requirements of engineers dealing with daily practical cases, and could be a significant improvement to existing design guides. The predictions of this proposed method show good agreement with the experimental results, especially for cases with three in-duct elements [51].

## 3.4 SUMMARY

Flow-generated noise is aerodynamic noise produced by in-duct elements on the quiet side of the primary attenuator in a ventilation system. The accurate prediction and control of flow-generated noise are important because, at long distances from the primary noise source (the fan), such noise from duct fittings could present severe problems.

This chapter reviewed studies on the measurement and prediction of flow-generated noise, particularly the prediction methods for flow-generated noise produced by in-duct discontinuities. To provide a basis for developing a generalised prediction method for the noise produced by all flow duct discontinuities in ventilation ductwork systems, several issues need to be addressed. (1) To accurately predict the flow-generated noise, the prediction method should consider the acoustics and aerodynamic interactions between flow duct elements. (2) To derive a generalised prediction technique, all of the data for a particular in-duct element or several similar elements must be collapsed onto a single curve. For different in-duct elements, a single normalised spectrum for flow-generated noise may be insufficient for all the configurations of in-duct elements. It is suggested that several representative normalised spectra should be obtained for different types of in-duct elements. (3) A simple method to determine the representative characteristic dimension of different flow discontinuities or in-duct elements is required to obtain the Strouhal number. In conclusion, it remains to be seen whether the developed prediction methods can be applied or extended to a wider configuration of obstructions in air duct systems. Some practical problems will be encountered in the prediction of flow-generated noise.

Assessing the effects of duct-borne sound and flow-generated noise on people is as important as their prediction. In Chapter 5, we will discuss the recent development of a holistic psychoacoustic assessment method for noise from ventilation systems.

## References

1. American Society of Heating, Refrigerating and Air-Conditioning Engineers, 2019 *ASHRAE Handbook*, HVAC Applications SI Edition.
2. Chartered Institution of Buildings Services Engineers (London), 2016 *CIBSE guide B4–Noise and vibration control for building services systems*, pp. 8–9.
3. T.K. Willson and A. Iqbal, 1980 *Building Services Engineering Research and Technology* 1, pp.54–57, Computer-aided analysis of airflow systems noise.
4. D.J. Croome and L.J. Stewart, 1971 *Journal of the Institution of Heating and Ventilating Engineers* 38, pp.239–251, Sound sources in airflow system.
5. C.G. Gordon, 1968 *The Journal of the Acoustical Society of America* 43, pp.1041–1048, Spoiler-generated flow noise. I. The experiment.
6. M.S. Howe and H.D. Baumann, 1992 *Noise and vibration control engineering, principles and applications*, Edited by L.L. Beranek and Istvan L.Ver, Chapter 14, pp.519–523, Noise of gas flows.
7. M.J. Lighthill, 1952 *Proceedings of the Royal Society of London. Series A. Mathematical and Physical Sciences* 211, pp.564–587, On sound generated aerodynamically I. General theory.
8. M.J. Lighthill, 1954 *Proceedings of the Royal Society of London. Series A. Mathematical and Physical Sciences* 222, pp.1–32, On sound generated aerodynamically II. Turbulence as a source of sound.
9. H.C. Hardy, 1963 *ASHRAE Journal* 5, p.95, Generalised theory for computing noise from turbulence and aerodynamic systems.
10. N. Curle, 1955 *Proceedings of the Royal Society of London. Series A. Mathematical and Physical Sciences* 231, pp.505–514, The influence of solid boundaries upon aerodynamic sound.

11. W.C. Meecham, 1965 *The Journal of the Acoustical Society of America* 37, pp.516–522, Surface and volume sound from boundary layers.
12. O.M. Phillips, 1956 *Proceedings of the Royal Society of London. Series A. Mathematical and Physical Sciences* 234, pp.327–335, On the aerodynamic surface sound from a plane turbulent boundary layer.
13. A. Powell, 1960 *The Journal of the Acoustical Society of America* 32, pp.982–990, Aerodynamic noise and the plane boundary.
14. H.G. Davies and J.E.F. Williams, 1968 *Journal of Fluid Mechanics* 32, pp.765–778, Aerodynamic sound generation in a pipe.
15. C.M. Mak, D.C. Waddington, and D.J. Oldham, 1997 *Building Acoustics* 4, pp.275–294, The prediction of airflow generated noise in ventilation systems.
16. G.K. Batchelor, 1951 *Mathematical Proceedings of the Cambridge Philosophical Society* 47, pp.359–374, Pressure fluctuations in isotropic turbulence.
17. I. Proudman, 1952 *Proceedings of the Royal Society of London. Series A. Mathematical and Physical Sciences* 214, pp.119–132, The generation of noise by isotropic turbulence.
18. A. Powell, 1964 *The Journal of the Acoustical Society of America* 36, pp.177–195, Theory of vortex sound.
19. U. Ingard, A. Oppenheim, and M. Hirschorn, 1968 *ASHRAE Transactions* 74, Noise generation in ducts.
20. W.W. Soroka, 1939 *Refrigeration Engineering* 37, p.393, Noise in ducts.
21. W.W. Soroka, 1970 *Applied Acoustics* 3, pp.309–321, Experimental study of high velocity air discharge noise from some ventilating ducts and elbows.
22. W.F. Kerka, 1960 *ASHRAE Journal* 2, p. 429, In high velocity system duct parts create sound as well as suppress it.
23. J.H. Watson, 1968 *Australian Refrigeration, Air Conditioning and Heating Journal* 22, pp.30–33, Acoustical characteristics of mitre bends with lined turning vanes.
24. H. Brockmeyer, July 1968, Dissertation for degree of Dr.-Ing. Carolo Wilhelmina, Technical University, Braunschweig, Flow acoustic study of duct fittings of high velocity air conditioning systems (H.V.R.A. Translation 195).
25. J.B. Chaddock, 1957 *Bolt Beranek and Newman Inc; Technical Information Report* No.45, Ceiling air diffuser noise.
26. M. Hubert, Feb/Apr 1969 *Larmbekompfung* 46–51, pp.29–33, Noise development in ventilation plant (H.V.R.A. Translation 162).
27. M. Hubert, 1968 *Akustiche Konferenz Budapest* IV, Beitrag, Gerausche durchstromter Gitter.
28. E.I. Iudin, 1955 *Soviet Physics Acoustics* 1, pp.383–398, The acoustic power of the noise created by airduct elements.
29. C.G. Gordon, 1969 *The Journal of the Acoustical Society of America* 45, pp.214–223, Spoiler-generated flow noise. II. Results.
30. C.G. Gordon, 1968 *ASHRAE Transaction* 2070, p. 13, The problem of duct generated noise and its prediction.
31. C.G. Gordon and G. Maidanik, 1966 *Bolt Beranek and Newman Inc. Report* No.1426, Influence of upstream flow discontinuities on the acoustic power radiated by a model air jet.

32. H.H. Heller and S.E. Widnall, 1970 *The Journal of the Acoustical Society of America* 47, pp.924–936, Sound radiation from rigid flow spoilers correlated with fluctuating forces.

33. H.H. Heller, S.E. Widnall, and C.G. Gordon, 1968 *Bolt Beranek and Newman Inc. Report* No.1734 Correlation of fluctuating forces with the sound radiation from rigid flow spoilers.

34. M.J. Holmes, 1973 *H.V.R.A. Lab. Report* No.78, Air flow generated noise Part I: Grilles and dampers.

35. M.J. Holmes, 1973 *H.V.R.A. Lab. Report* No.78, Air flow generated noise Part II: Bends with turning vanes.

36. P.A. Nelson and C.L. Morfey, 1981 *Journal of Sound and Vibration* 79, pp.263–289, Aerodynamic sound production in low speed flow ducts.

37. D.J. Oldham and A.U. Ukpoho, 1990 *Journal of Sound and Vibration* 140, pp.259–272, A pressure-based technique for predicting regenerated noise levels in ventilation systems.

38. A.U. Ukpoho and D.J. Oldham, 1991 *Proceedings of Institute of Acoustics* 13, pp.461–468, Regenerated noise levels due to closely spaced duct elements.

39. C.M. Mak and J. Yang, 2000 *Journal of Sound and Vibration* 229, pp.743–753, A prediction method for aerodynamic sound produced by closely spaced elements in air ducts.

40. C.M. Mak and J. Yang, 2002 *Acta Acustica united with Acustica, The journal of the European Acoustics Association (EEAA)* 88, pp.861–868, Flow-generated noise radiated by the interaction of two strip spoilers in low speed flow ducts.

41. C.M. Mak, 2002 *Applied Acoustics* 63, pp.81–93, Development of a prediction method for flow-generated noise produced by duct elements in ventilation systems.

42. C.M. Mak, 2005 *Journal of Sound and Vibration* 287, pp.395–403, A prediction method for aerodynamic sound produced by multiple elements in air ducts.

43. N. Han, X.J. Qiu, and C.M. Mak, 2006 *Journal of Sound and Vibration* 294, pp.374–380, A further study of the prediction method for aerodynamic sound produced by two in-duct elements.

44. N. Han and C.M. Mak, 2008 *Applied Acoustics* 69, pp.566–573, Prediction of flow-generated noise produced by acoustic and aerodynamic interactions of multiple in-duct elements.

45. C.M. Mak, J. Wu, C. Ye, and J. Yang, 2009 *The Journal of the Acoustical Society of America* 125, pp.3756–3765, Flow noise from spoilers in ducts.

46. C.M. Mak and D.J. Oldham, 1998 *Building Acoustics* 5, pp.123–141, The application of computational fluid dynamics to the prediction of flow generated noise in low speed ducts. Part 1: Fluctuating drag forces on a flow spoiler.

47. C.M. Mak and D.J. Oldham, 1998 *Building Acoustics* 5, pp.201–215, The application of computational fluid dynamics to the prediction of flow generated noise: Part 2: Turbulence-based prediction technique.

48. C.M. Mak and W.M. Au, 2009 *Applied Acoustics* 70, pp.11–20, A turbulence-based prediction technique for flow-generated noise produced by in-duct elements in a ventilation system.

49. C.M. Mak, X. Wang, and Z.T. Ai, 2014 *Applied Acoustics* 76, pp.386–390, Prediction of flow noise from in-duct spoilers using computational fluid dynamics.

50. C. Bullock, 1970 *ASHRAE Journal* 12, pp.39, Aerodynamic sound generation by duct elements.

51. C.Z. Cai and C.M. Mak, 2018 *Applied Acoustics* 135, pp.136–141, Generalized flow-generated noise prediction method for multiple elements in air ducts.

# Chapter 4

# Assessment of vibration isolation for machinery of ventilation systems

## 4.1 INTRODUCTION

Despite their benefits, ventilation ductwork systems produce airborne, duct-borne, and structure-borne sounds. Sounds are generated because the systems have vibrating mechanical machines, such as ventilation fans and motor-driven equipment. These machines may cause the roof or floor of the electrical and mechanical (E&M) equipment room to vibrate sufficiently high, rendering the spaces they serve unusable. Thus, satisfactory acoustic design plays an important role in the design of ventilation systems. Building acoustics is focused on the audio frequency range (or audible frequency range) 20–20 kHz as well as the octave or 1/3 octave scale and Bark scale for the frequency analysis in acoustic and psychoacoustic measurements, respectively. However, vibration or structure-borne sound analyses are typically centred on low frequencies (e.g., 0–200 Hz). This range includes frequencies from 20 to 200 Hz that humans may be able to hear and vibrations below 20 Hz that can be felt but are inaudible. The human perception of vibration and sound differs because our senses include hearing and touch. In addition, vibrations or structure-borne sound problems are typically analysed in a single-hertz band instead of the octave or 1/3 octave band.

Figure 4.1 shows a schematic of a simple vibratory system for modelling a machine resting on a floor/roof. It includes a simple vibration isolator in relation to a vibrating source, structure-borne sound transmission path, and receiver. It also demonstrates several important aspects in the study of structure-borne sound transmission in buildings such as:

- Structure-borne sound source characterisation,
- Vibration isolation (assessment of performance using transmissibility),
- Vibrational waves produced in the floor/roof, and
- Coupling with air at the hearing end.

DOI: 10.1201/9781003201168-4

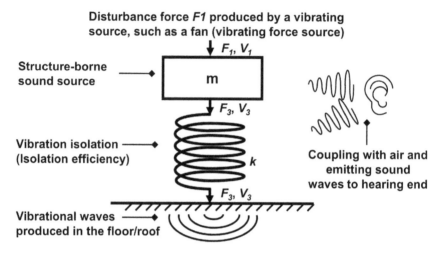

*Figure 4.1* A schematic of a simple vibratory system consisting of a mass *m* and a spring of stiffness *k* with a single contact point on a floor.

This chapter focuses on vibration isolation. First, the fundamentals of vibration and some terminologies are introduced in relation to source characterisation and vibration isolation.

## 4.2 FUNDAMENTALS OF VIBRATION

Vibrations can be classified as periodic, random, or transient. In *periodic vibrations*, motion is exactly repeated after a time interval called period. The simplest type of periodic vibration, called *simple harmonic motion*, can be presented as a displacement–time $(x–t)$ graph or sine wave. This type of motion can be described in terms of a single frequency $f$. In *random vibrations*, oscillations are never exactly repeated. An example is the vibration produced by wind in building structures. *Transient vibrations* diminish to zero after a certain period. An example is the vibration in buildings generated by heavy vehicle passage.

### 4.2.1 Simple oscillator (free undamped vibration)

When mass $m$, fastened to a spring and constrained to move parallel to the spring, is slightly displaced from its rest position and released, the mass vibrates (Figure 4.2). The sinusoidal vibrations of this mass are known as *simple harmonic vibrations*.

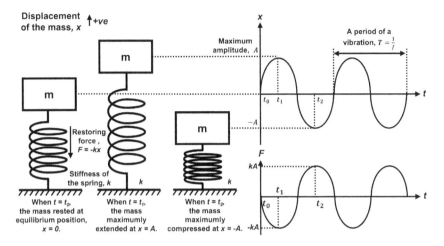

*Figure 4.2* Simple oscillator.

As shown in Figure 4.2, the restoring force, $F$, can be given by:

$$F=-kx,$$ (4.1)

where $x$ is the displacement of mass, and $k$ is the spring stiffness.

The substitution of this force expression into the general equation of linear motion yields

$$F=m\frac{d^2x}{dt^2}.$$ (4.2)

A linear differential equation can be obtained as follows:

$$\frac{d^2x}{dt^2}+\omega_n^2x=0,$$ (4.3)

where $\omega_n^2=k/m$.

The complete general solution is

$$x=A_1\cos\omega_n t+A_2\sin\omega_n t,$$ (4.4)

where $A_1$ and $A_2$ are arbitrary constants; $\omega_n$ is the angular frequency in radians per second (rad/s); and the natural frequency is $f_n=\omega_n/2\pi$.

## 4.2.2 Initial conditions

The differentiation of Eq. 4.4 and substitution of initial speed at $t = 0$ yields $u_o = \omega_n A_2$. The following form is derived:

$$x = x_o \cos \omega_n t + \frac{u_o}{\omega_n} \sin \omega_n t. \tag{4.5}$$

Let $A_1 = A\cos\varphi$ and $A_2 = -A\sin\varphi$, where $A$ and $\varphi$ are new arbitrary constants. By substituting these expressions, another form of Eq. 4.4 may be obtained:

$$x = A\cos(\omega_n t + \varphi), \tag{4.6}$$

where $A$ is the amplitude, and $\phi$ is the initial phase angle of motion. The initial conditions determine their values:

$$A = \sqrt{x_o^2 + \left(\frac{u_o}{\omega_n}\right)^2} \quad \text{and} \quad \varphi = \tan^{-1}\left(-\frac{u_o}{\omega_n x_o}\right). \tag{4.7}$$

## 4.2.3 Damped oscillations (free vibration)

Whenever a real body oscillates, dissipative (frictional) forces are generated. These forces are of many types, depending on the oscillating system. However, they consistently result in the damping of oscillations—a decrease in the amplitude of free oscillations with time (Figure 4.3(b)).

Consider the viscous frictional force, $F_r$, on a simple oscillator. Such a force is assumed to be proportional to the speed of mass and directed to oppose the motion. It can be expressed as

$$F_r = -R_m \frac{dx}{dt}, \tag{4.8}$$

where $R_m$ is a positive constant known as the mechanical resistance of the system. In this case of viscous damping, the amount of damping is quantified using the damping constant, $R_m$. This is infrequently denoted by another symbol, $C$, defined as the damping force per unit velocity.

The equation of motion of the oscillator constrained by the stiffness force $(-kx)$ becomes

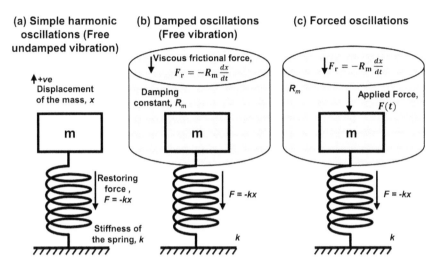

Figure 4.3 Oscillators with (a) free undamped vibration; (b) free vibration; and (c) forced vibration.

$$\frac{d^2x}{dt^2} + \frac{R_m}{m}\frac{dx}{dt} + \omega_n^2 x = 0, \tag{4.9}$$

where $\omega_n^2 = k/m$.

The general solution of this differential equation is

$$x = Ae^{-\beta t}\cos(\omega_d t + \varphi), \tag{4.10}$$

where $A$ and $\varphi$ are real constants determined by the initial conditions; $\beta = R_m/2m$; and the angular frequency of the damped oscillator is $\omega_d = \sqrt{\omega_n^2 - \beta^2}$.

## 4.2.4 Forced oscillations

A simple oscillator is typically driven by an externally applied force, $f(t)$, Figure 4.3(c). The differential equation for motion becomes

$$m\frac{d^2x}{dt^2} + R_m\frac{dx}{dt} + kx = F(t). \tag{4.11}$$

If the periodic driving force with the driving angular frequency, $\omega$, is $F(t) = F\cos \omega t$, then the above equation becomes

$$m\frac{d^2 x}{dt^2} + R_m \frac{dx}{dt} + kx = F\cos \omega t. \tag{4.12}$$

The solution of the differential equation is

$$x = Ae^{-\beta t}\cos(\omega_d t + \varphi) + (F/\omega Z_m)\sin(\omega t - \theta), \tag{4.13}$$

where $Z_m$ is the complex mechanical impedance of the system. It is given by $Z_m = R_m + jX_m$ in which the mechanical reactance is $X_m = \omega m - \dfrac{k}{\omega}$, and $\theta$ is given by $\theta = \tan^{-1}\dfrac{\omega m - k/\omega}{R_m}$.

## 4.3 SOURCE CHARACTERISATION

### 4.3.1 Introduction

Structure-borne sound problems in buildings are more difficult to solve than airborne or duct-borne sound problems. One reason is that no consensus on how these machines can be described as vibrational sources has been achieved. Several studies have demonstrated [1–5] that the activities and dynamic characteristics of a machine are both required to describe appropriately the ability of a vibrating source to emit structure-borne sound.

This section briefly introduces some basic terminologies and methods for characterising the source and receiver for evaluating power transmission.

### 4.3.2 Basic theory

#### 4.3.2.1 Mobility

The behaviour of most structures in buildings is essentially linear. This allows their dynamic response to be described by two basic parameters: exciting force and response velocity.

In mechanical vibrations, the *complex* ratio of the translational or rotational response velocity taken at a point in a system to the exciting force or moment phasor at the same point or another point in a system is defined as mobility. During excitation, all other points in the structure are allowed to respond freely without any constraint other than those representing the normal support of the structure. Following the foregoing general definition, other terms, such as point, transfer, and cross mobilities designating a location and direction with respect to two phasors, can also be defined [1–5].

Mobility is <u>frequency-dependent</u>. It is a function of structural geometry, boundary conditions, material properties, and damping. Analytical expressions for beams and infinite plates are available; however, publications on plate structures with finite boundaries are limited.

### 4.3.2.2 Free velocity

Machines, such as fans, generate vibrations during operation owing to internal dynamic forces. These forces are difficult to measure directly because a machine may have multiple vibration-generating components. The most practical way to describe the strength of a vibration source is to adopt a collective response approach. In this method, the complex source mechanism and internal transmission paths are represented by free velocities at contact points.

Free velocity (the velocity of a source at a contact point) is measured while the machine operates without contact with supporting structures.

### 4.3.2.3 Vibrational power transmission

For a vertical motion system with a single contact point, assuming harmonic time dependence, the complex time-averaged power can be expressed as [6]:

$$P = \frac{1}{2} F^* v, \tag{4.14}$$

where $F$ and $v$ are the force and velocity at the contact point, respectively; the asterisk denotes a complex conjugate.

The power transmitted, expressed in terms of the source and receiver properties, can be expressed as [7]:

$$P = \frac{1}{2} \frac{\left| v_{sf}^{2} \right|}{\left| Y_s + Y_r \right|^2} Re(Y_r), \tag{4.15}$$

where $v_{sf}$ is the root mean square free velocity of the source, and the complex source and receiver mobilities are $Y_s$ and $Y_r$, respectively.

### 4.3.2.4 Source descriptor

Mondot and Petersson [7] proposed a novel characterisation of structure-borne sound sources in terms of a source descriptor describing the internal activity and dynamic characteristics of a source using the free velocity and mobility at contact points. By introducing $Y_s^*$ (complex conjugate of source mobility) into the numerator and denominator of Eq. 4.15, the structure-borne sound source can be characterised. Accordingly, the complex power is given by

$$P = \frac{1}{2} \frac{|v_{sf}|^2}{Y_s^*} \frac{Y_s^* Y_r}{|Y_s + Y_r|^2}. \qquad (4.16)$$

The first term, $\dfrac{1}{2} \dfrac{|v_{sf}|^2}{Y_s^*}$, called the *source descriptor*, is solely a function of the sound source. Although its dimension is that of power, it is not actual power; it represents the ability of the source to deliver power. The actual power transmitted is a fraction of this source descriptor. It is determined by the second term, the *coupling function*, which is the degree of matching between the source and receiver. The source descriptor data can be obtained prior to installation; thus, this unique and invariant quantity is suitable for characterising the sound source based on unifying power. This concept was subsequently extended to studies of multipoint and multicomponent sources [2, 5] using effective mobility [8].

### 4.3.3 Summary

To evaluate the structure-borne sound transmission from a machine to its supporting structure, characterising the structure-borne sound source is necessary. One proposed method for source characterisation requires three basic parameters: free velocity, source mobility, and supporting structure mobility. The use of a single parameter is insufficient for properly characterising the source or predicting the energy flow under the installation condition of the machine.

### 4.4 VIBRATION ISOLATION

### 4.4.1 Introduction

Air-cooled chillers, chilled water pumps, and primary air handling units are typically installed on the E&M roof or floor of a building. The *ASHRAE Handbook* [9] recommends that a noise-sensitive area must not be near or adjacent to E&M plant rooms. However, for noise-sensitive rooms (e.g., penthouse-type executive offices) not to be near or directly under the E&M floor on which air conditioning equipment (including air-cooled chillers or ventilation fans) is installed is unavoidable. Structure-borne sound problems are caused by machine vibrations. Although solving this problem is important, no simple or universally agreed upon method for characterising the 'noisiness' of equipment (such as compressors as a source of structure-borne sound) has been devised [1–5]. Several investigators [8, 10, 11] have endeavoured to devise techniques for predicting the occurrence of structure-borne sound. Nevertheless, no practical method has been developed for engineers to predict the level of structure-borne sound generated by machinery in occupied spaces. However, structure-borne sound can be reduced by installing vibration isolators. The selection of vibration isolators is based on the disturbance frequency, which is typically relatively low. Force transmissibility or isolation efficiency is generally adopted in the industry as a basis for selecting vibration isolators. In defining force transmissibility, the floor is assumed as non-movable (i.e., the floor mobility is zero). This means that the effect of floor mobility on isolation efficiency is generally ignored. In some cases where floor mobility was ignored, no structure-borne sound problems occurred. However, in certain cases, a considerable amount of vibrational energy was possibly transmitted to the floor. Consequently, vibrations were generated on the E&M floor and emitted noise to the room just below the floor. This possibly occurred because the effect of floor mobility was ignored in the usual force transmissibility and selection of vibration isolators. The effect of floor mobility on the isolation efficiency of vibration isolators, recently studied by Mak and Su [12], is introduced in this section.

## 4.4.2 Force transmissibility and isolation efficiency

### 4.4.2.1 Usual engineering definition of isolation efficiency and transmissibility

Machine vibrations are complicated. In practice, a simple single-degree-of-freedom undamped system with a mass and spring isolator is considered. By referring to Figure 4.4, the (force) transmissibility [13] for an undamped system can be defined as follows:

$$T = \left| \frac{F_T}{F_o} \right| = \frac{1}{\left| 1 - \left( \dfrac{f_d}{f_n} \right)^2 \right|} = \frac{1}{\left| 1 - \left( \dfrac{\omega}{\omega_n} \right)^2 \right|}, \tag{4.17}$$

where
$F_T$ = transmitted force through the spring into the surface of the floor;
$F_o$ = source force;
$f_d$ = forcing frequency;
$f_n$ = natural frequency;
$\omega$ = angular forcing frequency; and
$\omega_n$ = angular natural frequency.

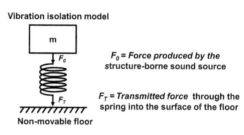

Vibration isolation model

$F_0$ = Force produced by the structure-borne sound source

$F_T$ = Transmitted force through the spring into the surface of the floor

Non-movable floor

*Figure 4.4* A vibration isolation model.

Isolation efficiency, $I$, is then defined as

$$I = 1 - T = 1 - \frac{1}{\left|1 - \left(\dfrac{f_d}{f_n}\right)^2\right|} = 1 - \frac{1}{\left|1 - \left(\dfrac{\omega}{\omega_n}\right)^2\right|}. \tag{4.18}$$

In the foregoing, $f_n$ is the natural frequency given by

$$f_n = \frac{1}{2\pi}\sqrt{\frac{k}{m}} = \frac{1}{2\pi}\sqrt{\frac{g}{\delta_{St}}} = 15.8\sqrt{\frac{1}{\delta}}\,(\text{Hz}), \tag{4.19}$$

where
$\delta_{st}$ = static deflection in metre = $mg/k_x$;
$\delta$ = static deflection in millimetre = $\delta_{st} \times 10^3$
(usually given in the manufacturer's catalogue);
$k$ = axial stiffness of the spring;
$g$ = acceleration due to gravity; and
$m$ = effective mass of the simple vibration isolation model.
In this formula, the floor is assumed to be non-movable.

Figure 4.5 shows a plot of the usual engineering transmissibility against $f_d/f_n$ for a single degree-of-freedom undamped system. The figure indicates that (1) when $f_d > \sqrt{2}\, f_n$, the transmissibility is less than unity, and useful isolation commences. (2) When $f_d < \sqrt{2}\, f_n$, force amplification occurs. (3) The higher the $f_d$ value, or the lower the $f_n$ value, the greater the degree of isolation. (4) When $f_d = f_n$, $T \rightarrow \infty$; however, this does not occur in practice because of the presence of damping or mechanical restraints/constraints within the system. In the foregoing (where the disturbance

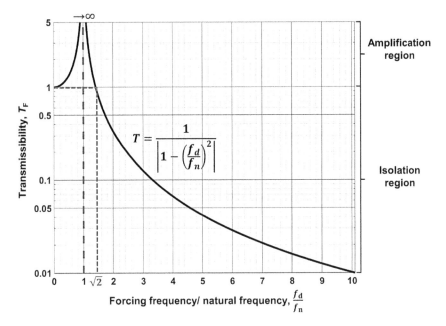

*Figure 4.5* Usual engineering transmissibility curve for an undamped system.

frequency equals the natural frequency), the situation is known as 'resonance condition'.

### 4.4.2.2 General practice in selecting vibration isolators

Various types of vibration isolators are shown in Figure 4.6. Ventilation system designers select the equipment based on various design criteria. From the manufacturer's catalogue the net weight (in kilogram) of a machine can be obtained. Typically, designers add approximately 20% to this weight as a safety factor. Based on the manufacturer's catalogue, a suitable static deflection of a vibration isolator can be selected by the contractor according to the weight of the machine provided by the consultant. The natural frequency can then be calculated from the static deflection (in millimetres) according to the usual definition of the natural frequency given by Eq. 4.19. Based on the calculated natural and disturbance frequencies, the isolation efficiency can be calculated and verified according to the typical definition given by Eqs. 4.16 and 4.17. The isolation efficiency is expected to exceed 90% ($f_d$/ $f_n$ > 3.3). The general rule in practice is that the greater the static deflection ($\delta$), the lower the natural frequency ($f_n$) of the system. This results in the

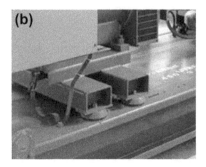

*Figure 4.6* Various types of vibration isolators (a) metal springs and (b) rubber mounts.

greater isolation efficiency of the spectrum of disturbing frequencies ($f_d$) in which isolation occurs.

Two types of vibration isolators are commonly used in Hong Kong.

1. Spring isolator ($\delta \geq 25$ mm)

This isolator typically has high static deflection such that it provides satisfactory vibration isolation even at a low disturbance frequency. At high disturbance frequencies, a bridging effect may occur and can be eliminated by incorporating a neoprene pad on the base plate. Typically, the installation cost is high.

2. Rubber (neoprene) ($\delta \leq 12$ mm)

The use of this rubber to isolate high disturbance frequencies is effective because of its high natural frequency. However, performance is degraded after the long exposure of rubber to sunlight. This isolator only requires simple installation and is typically inexpensive.

In theory, the ratio of the forcing frequency to the natural frequency ($f_d/f_n$) must be as large as possible provided that it is compatible with the stability. A vibratory machine can be rigidly mounted onto an *inertia block* (a large concrete block) and separated by vibration isolators. This allows the stiffer isolators to achieve the same natural frequency. In addition, the assembly's centre of gravity was lowered, and the stability improved. A study on the effect of the inertia block on the stability of the vibratory system and performance of vibration isolation can be found in [14].

### 4.4.2.3 Real isolation efficiency and transmissibility

Different from their usual definitions, transmissibility and isolation efficiency were described by Mak and Su [12] by considering the effect of floor mobility. For the four identical simple isolators with four contact points symmetric about the central point of a plate, as shown in Figure 4.7, 'real' transmissibility (referred to as *'real'* here because the exhibited transmissibility is believed to reflect reality) is expressed by the following [12]:

$$T = \left| \frac{F_i'}{F_i} \right| = \frac{1}{\left| 1 - \dfrac{\omega^2}{\omega_n^2} + j\omega m (Y_i) \right|} \quad for \quad i = 1, 2, 3, 4, \qquad (4.20)$$

where

$Y_i = Y_{i1} + Y_{i2} + Y_{i3} + Y_{i4}$ = effective floor mobility of contact
point $i$ ($i$ = 1, 2, 3, 4)
(defined in [8]);
$F_i'$ = transmitted force for $i^{th}$ isolator ($i$ = 1, 2, 3, 4); and
$F_i$ = source force for $i^{th}$ isolator ($i$ = 1, 2, 3, 4).
If $Y_{ij}$ = 0 for $i$ and $j$ = 1, 2, 3, 4 (i.e., the floor is non-movable), then Eq. 4.20 is transformed into the usual definition of transmissibility (i.e., Eq. 4.17).

The isolation efficiency, $I$, is given by the following [12]:

$$I = 1 - T = 1 - \left| \frac{F_i'}{F_i} \right| = 1 - \frac{1}{\left| 1 - \dfrac{\omega^2}{\omega_n^2} + j\omega m (Y) \right|} \quad (for \quad i = 1, 2, 3, 4). \qquad (4.21)$$

### 4.4.2.4 Comparison between 'usual' and 'real' transmissibilities

A vibratory machine placed on a floor is modelled in Figure 4.7. Four simple symmetric isolators (each with simple mass ($m$) and spring stiffness ($k$)) are placed symmetrically on a simple concrete plate. The physical parameters of the plate are density ($\rho$= 2.8 × 10³ kg/m³), Young's modulus ($E$ = 2.1 × 10¹⁰ N/m²), loss factor ($\eta$= 0.5 × 10⁻²), and Poisson's Ratio ($\mu$ = 0.2). The geometric dimensions of the plate are as follows: (length) $l$ = 3.5 m, (width) $W$ = 3.5 m, and (thickness) $d$ = 0.24 m.

The floor mobility, Y, of the contact point can then be obtained, as shown in Figure 4.8a. If the mass, $m$, is 400 kg and the disturbance frequency, $f_d$, is 50 Hz, the transmissibility curves can be obtained using Eqs. 4.16 and 4.19. In Figure 4.8b, the dotted line represents the transmissibility obtained by the usual engineering definition (i.e., Eq. 4.1), and the solid line represents the real transmissibility described in this study [12]. In the real transmissibility, the plate is assumed to have a vibration contact point, whereas in the usual engineering transmissibility, no contact

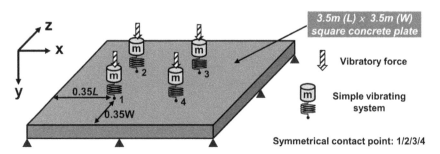

*Figure 4.7* A schematic of simple vibrating systems with four symmetrical contact points placed symmetrically on a simply-simply-simply-simply plate.

occurs (i.e., $Y = 0$). The real transmissibility curve is based on the value of $Y$ in Figure 4.8a. By contrast, $Y = 0$ is assumed in the typical engineering transmissibility. Figure 4.8b indicates that the two transmissibility curves have considerable differences. Real transmissibility is larger than usual transmissibility when the ratio of the forcing frequency to the natural frequency ($f_d/f_n$) exceeds 2.6. The real transmissibility curve reaches its peak when $f_d/f_n$ is approximately 3.5. In other words, the real isolation efficiency is smaller than the usual isolation efficiency when $f_d/f_n$ is greater than 2.6, and it reaches its lowest value when $f_d/f_n$ is near 3.5. The common presumption is that $f_d/f_n > 3$ ($T < 13\%$) can be used to select vibration isolators. However, Figure 4.8b shows that the real transmissibility can be considerably higher than the usual engineering transmissibility when $f_d/f_n > 3$ (particularly $f_d/f_n = 3.6$). The usual transmissibility is only 8.3%; in contrast, the real transmissibility exceeds 70%. The difference between the two in Figure 4.8b decreases as $f_d/f_n$ increases. If $f_d/f_n$ is large, the effect of the floor mobility, $Y$, of the plate can be ignored. When $f_d/f_n$ is approximately 3.5, the real transmissibility is significantly greater than the usual transmissibility. Consequently, the actual isolation efficiency obtained using Eq. 4.21 is considerably less than the usual isolation efficiency obtained using Eq. 4.18. This possibly explains why vibration isolators sometimes do not perform well although they are selected according to the usual isolation efficiency.

The force transmissibility method, which is based on the usual engineering definition of force transmissibility from Eq. 4.17, ignores the effect of floor mobility. The technique is based on an undamped lumped parameter system (i.e., a long wavelength limit) with a single degree of freedom (i.e., only vertical linear motion). In addition, a single contact point ignores the interactions of all the dynamic forces transmitted to the floor from the machine through the vibration isolators.

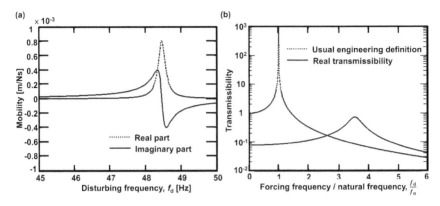

*Figure 4.8* (a) Mobility of the simply-simply-simply-simply plate against frequency and (b) transmissibility curves.

### 4.4.2.5 Damped forced oscillatory system

The effect of damping is ignored in Eq. 4.17. However, in all linked systems, damping is present to varying degrees; consequently, it modifies some of the relationships.

The force transmissibility, $T_F$, for a viscously damped forced oscillatory system is expressed as follows:

$$T_F = \sqrt{\frac{1 + 4\xi^2 \left(\frac{f_d}{f_n}\right)^2}{(1 - \left(\frac{f_d}{f_n}\right)^2)^2 + 4\xi^2 \left(\frac{f_d}{f_n}\right)^2}}, \tag{4.22}$$

where
$f_d$ = forcing frequency;
$f_n$ = natural frequency;
$\xi = C/C_o$, damping ratio of the mass–spring system;
$C$ = actual damping; and
$C_o$ = critical damping (the damping value at which the oscillating motion commences).

An examination of the family of curves in Figure 4.9 shows that increasing the value of damping reduces not only the peak transmissibility at resonance but also the performance of the vibration isolator in the useful isolation region. To achieve maximum isolation, damping must be as low as possible; this is acceptable for a majority of applications encountered by building service engineers.

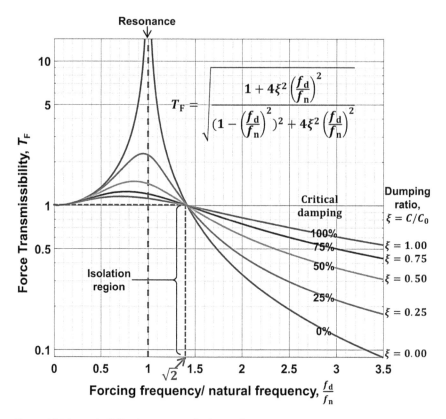

Figure 4.9 Transmissibility for a viscously damped system.

### 4.4.3 Power transmissibility

The force transmissibility method is commonly adopted in the industry because it is simple and convenient. However, it neglects the effect of floor mobility on the structure-borne sound power transmitted from a vibratory machine to the floor/roof as well as the interactions among several contact points between the vibratory machine and floor/roof. This is because the method is based on a single contact point, single-degree-of-freedom (vertical linear motion), and lumped parameter system. Actual vibrating machines, such as fans, typically have several mounting points. The use of the power transmissibility method to assess the performance of vibration isolators may not be correct considering the following. (1) The effect of floor mobility cannot be ignored at natural frequencies. (2) The weight of a vibrating machine is not evenly distributed among several mounting points, or (3) other vibrations exist in other degrees of freedom (such as rotations) at one of the mounting points.

The '*power transmissibility method*' proposed by Mak et al. [15–18] was used to evaluate the vibration–isolation performance of a machine in a building. It is based on the ratio of structure-borne sound power. The power transmissibility is $\gamma = \dfrac{P_{t(s)}}{P_{t(ns)}}$, where $P_{t(ns)}$ is the active structure-borne sound power transmitted to the floor from a vibratory machine mounted without a vibration isolator. Moreover, $P_{t(s)}$ is the active structure-borne sound power transmitted to the floor from the machine mounted with a vibration isolator.

For a vibratory system with one contact point, the power transmissibility is given by

$$\gamma = \frac{P_{exact(s)}}{P_{exact(ns)}} = \frac{\left|1 + j\omega Y_i\right|^2}{\left|1 - \left(\dfrac{\omega}{\omega_n}\right)^2 + j\omega m Y_i\right|^2}. \tag{4.23}$$

The floor mobility, $Y_r$, is extremely small such that $\left|j\omega m Y_r\right| \ll 1$, and $\left|j\omega m Y_r\right| \ll \left|1 - (\omega/\omega_n)^2\right|$. Therefore,

$$\gamma \approx \left|\frac{1}{1 - (\omega/\omega_n)^2}\right|^2 = T^2.$$

For a vibratory system with four contact points, the power transmissibility is given by

$$\gamma = \frac{\begin{bmatrix} 1 & 1 & 1 & 1 \end{bmatrix}\left(\left[Y_a\right] - \left(\dfrac{2\omega}{\omega_n}\right)^2 [I] + j\omega M\left[Y_r\right]\right)^{-1T^*}\left(\mathrm{Re}\left(\left[Y_r\right]\right)\right)}{\begin{bmatrix} 1 & 1 & 1 & 1 \end{bmatrix}\left(\left[Y_a\right] + j\omega M\left[Y_r\right]\right)^{-1T^*}\mathrm{Re}\left(\left[Y_r\right]\right)} \\ \times \left(\left[Y_a\right] - \left(\dfrac{2\omega}{\omega_n}\right)^2 [I] + j\omega M\left[Y_r\right]\right)^{-1}\begin{bmatrix} 1 \\ 1 \\ 1 \\ 1 \end{bmatrix}}{\quad\quad\quad\quad\quad \times \left(\left[Y_a\right] + j\omega M\left[Y_r\right]\right)^{-1}\begin{bmatrix} 1 \\ 1 \\ 1 \\ 1 \end{bmatrix}}. \tag{4.24}$$

This expression for $\gamma$ is more complicated than that for a system with a single contact point.

Because the floor mobility matrix, $[Yr]$, is extremely small, $j\omega M[Y_r] \ll [Y_a] - \left(\dfrac{2\omega}{\omega_n}\right)^2 [I]$, and $j\omega M[Y_r] \ll [Y_a]$, the power transmissibility becomes

$$\gamma \approx \dfrac{\dfrac{1}{16}|F_o|^2 [1 \quad 1 \quad 1 \quad 1] \left(\dfrac{1}{1-\left(\dfrac{\omega}{\omega_n}\right)^2}\right)^2 \mathrm{Re}\left([Y_r]\right)\begin{bmatrix}1\\1\\1\\1\end{bmatrix}}{\dfrac{1}{16}|F_o|^2 [1 \quad 1 \quad 1 \quad 1]\mathrm{Re}\left([Y_r]\right)\begin{bmatrix}1\\1\\1\\1\end{bmatrix}} = \left(\dfrac{1}{1-\left(\dfrac{\omega}{\omega_n}\right)^2}\right)^2 = T^2.$$

(4.25)

The advantage of the power transmissibility method is that it considers the effect of floor mobility and interaction among all the dynamic forces transmitted to the floor through vibration isolators. Further studies [19] were conducted to incorporate the effect of viscous damping into the power transmissibility method. In addition, recent studies [14, 20] have been conducted to investigate the effect of inertia block on the stability of the vibratory system and performance of vibration isolation. Further work on the isolation of transient vibrations was conducted by Wang and Mak [21]. Therefore, the performance of vibration isolation can be assessed using the power transmissibility method. However, the method is constrained because it requires the availability or measurements of the source mobility, floor mobility, and the free velocity of the machine (vibratory source). The source mobility can be measured at the same support points using force actuators mounted with the elastically suspended machine. The floor mobility can be measured using electrodynamic shakers and force actuators. Mobility is measured as a transfer function using a dual-channel fast Fourier transform analyser.

## 4.4.4 Summary

The effect of floor mobility on the force transmissibility and isolation efficiency of vibration isolators is discussed. The effective floor mobility increases the force transmissibility. In turn, the isolation efficiency decreases when $f_d/f_n$ exceeds 2.6 for the four contact points on the plate [16]. Other practical problems concerning vibration isolators are found in [13]. Floor mobility is important in the selection of vibration isolators because it affects the isolation efficiency. The presumption [16] is that the sound radiating from the E&M equipment floor above a room and the walls surrounding the room can be reduced if floor mobility is considered in the selection of vibration isolators. The developed power transmissibility method [15] can estimate the performance of vibration isolators because it considers the effect of the dynamic interaction among all contact points and floor mobility.

## 4.5  USE OF INERTIA BLOCK FOR STABILISATION OF VIBRATION-ISOLATED MACHINES

The inertia block shown in Figure 4.10 is commonly used by engineers to stabilise vibration-isolated vibratory equipment. However, no comprehensive guidelines for the selection of inertia blocks or indices for assessing the effect of these blocks on the stability of vibratory systems are available in the building industry. In addition to vibration isolation, the purpose of maintaining the 'stability' of the mounted vibratory system is to control the vibration and rotational velocities of mounted machines. This can prevent the vibrating and rocking motions from adversely affecting the mounting and normal operations of machines. In other words, the use of a heavy and rigid inertia block can reduce the motion of an isolated machine. Yun and Mak [14] proposed the use of 'mounted vibration velocity' and 'mounted rotational velocity level' of the vibratory machine for assessing the stability of the vibratory system. To evaluate the effect of the inertia block on the performance of the vibration isolation system, a vibratory machine model with four mounting points was used and analysed. The mass distribution was uneven, and the exciting force was eccentric. They indicated that the use of an inertia block did not affect the vibration isolation performance. Instead, the block decreased the vibration velocity and rotational velocity of the isolated vibratory machine. As a result, the inertia block increased the stability of the vibratory system regardless of whether the machine had a slightly or highly uneven mass distribution. Their results revealed that, for

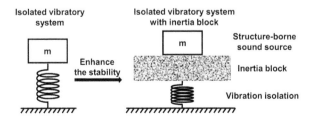

*Figure 4.10* An isolated machine with an inertia block.

a machine with a highly uneven mass distribution, the mounted vibratory machine required an inertia block with a large mass to enhance the stability of the isolated vibratory system.

## 4.6 VIBRATIONAL WAVES GENERATED IN SOLID FLOOR

The three main types of waves formed in solids are *longitudinal, transverse,* and *bending.* In general, longitudinal waves do not cause considerable direct radiation of sound into air. Nevertheless, they are important because of their ability to excite other parts of a structure to generate bending vibrations. These bending waves are extremely important because of their ability to radiate sounds. Further, they produce large deflections in the structure and cause considerable radiation of sound energy.

## 4.7 COUPLING WITH AIR AND EMISSION OF SOUND

Owing to the occasional inadequate design of vibration isolation, vibrations may be produced at the E&M floor that may, in turn, emit noise to the room just below this floor.

## 4.8 SUMMARY

This chapter introduces some fundamental knowledge on vibration isolation and the assessment of the vibration isolation performance of air conditioning or ventilation equipment. To understand the effect of acoustic environment on people due to vibration or sound sources, the succeeding chapter discusses psychoacoustic approaches.

## References

1. B.M. Gibbs, B.A.T. Petersson, and S.Y. Qiu, 1991 *Noise Control Engineering Journal* 37, pp.53–61, The characterization of structure-borne emission of building services machinery using the source descriptor concept.
2. B.A.T. Petersson and B.M. Gibbs, 1993 *Journal of Sound and Vibration* 168, pp.157–176, Use of the source descriptor concept in studies of multi-point and multi-directional vibrational sources.
3. J.X. Su, A.T. Moorhouse, and B. Gibbs, 1995 *Journal of Sound and vibration* 185, pp.737–741, Towards a practical characterization for structure-borne sound sources based on mobility techniques.
4. A. Moorhouse, 1995 *Building Services Engineering Research & Technology* 16, pp.B9–B12, Characterisation of sources of structure-borne sound.
5. R.A. Fulford and B.M. Gibbs, 1997 *Journal of Sound and Vibration* 204, pp.659–677, Structure-borne sound power and source characterisation in multi-point-connected systems, part 1: Case studies for assumed force distributions.

6. L. Cremer and M. Heckl *Structure-borne sound*, 1973. Springer-Verlag: New York.

7. J.M. Mondot and B.A.T. Petersson, 1987 *Journal of Sound and Vibration* 114, pp.507–518, Characterization of structure-borne sound sources: the source descriptor and the coupling function.

8. B. Petersson and J. Plunt, 1981 *Journal of Sound and Vibration* 82, pp.517–529, On effective mobilities in the prediction of structure-borne sound transmission between a source structure and a receiving structure, part I: theoretical background and basic experimental studies.

9. American Society of Heating, Refrigerating and Air-Conditioning Engineers, 1976 *ASHRAE Handbook & Product Directory*, Systems Volume.

10. B. Petersson and J. Plunt, 1981 *Journal of Sound and Vibration* 82, pp.517–529, On effective mobilities in the prediction of structure-borne sound transmission between a source structure and a receiving structure, part II: Procedures for the estimation of mobilities.

11. A. Moorhouse and B. Gibbs, 1993 *Journal of Sound and Vibration* 167, pp.223–237, Prediction of the structure-borne noise emission of a machines: development of a methodology.

12. C.M. Mak and J.X. Su, 2001 *Journal of Low Frequency Noise, Vibration and Active Control* 20, pp.1–13, A study of the effect of floor mobility on isolation efficiency of vibration isolators.

13. A. Fry, 1988 *Noise Control in Building Services*, edited by Alan Fry, Chapter 6, Vibration isolation, Sound Research Laboratories Ltd., Pergamon Press, pp. 131–185.

14. Y. Yun, C.M. Mak, and S.K. Tang, 2007 *Applied Acoustics* 68, pp.1511–1524, A study of the effect of inertia blocks on the stability of the vibratory system and the performance of vibration isolation.

15. C.M. Mak and J.X. Su, 2002 *Applied Acoustics* 63, pp.1281–1299, A power transmissibility method for assessing the performance of vibration isolation of building services equipment.

16. C.M. Mak and J.X. Su, 2003 *Building and Environment* 38, pp.443–455, A study of the effect of floor mobility on structure-borne sound power transmission.

17. C.M. Mak and G.K.C. Or, 2003 *Architectural Science Review* 46, pp.193–205, Development of an insertion loss for vibration isolation of building services equipment.

18. C.M. Mak and C.P. Tse, 2004 *Building Services Engineering Research and Technology* 25, pp.159–167, A performance indicator for vibration isolation of building services equipment.

19. J.C. Tao and C.M. Mak, 2006 *Applied Acoustics* 67, pp.733–742, Effect of viscous damping on power transmissibility for the vibration isolation of building services equipment.

20. Y. Yun and C.M. Mak, 2010 *Building and Environment* 45, pp.758–765, Assessment of the stability of isolated vibratory building services systems and the use of inertia blocks.

21. J.F. Wang and C.M. Mak, 2013 *Journal of Vibration and Control* 19, pp.2459–2468, An indicator for the assessment of isolation performance of transient vibration.

# Chapter 5

# Development of psychoacoustic prediction method

## 5.1 INTRODUCTION

The previous chapters discuss how duct-borne, flow-generated, and vibration noise is produced in ventilation ductwork systems. Over the last two decades, awareness of the negative impact of noise on the environment has increased [1]. The amount of scientific evidence on the impact of noise on aspects, such as working productivity, comfort, health, and well-being of occupants, continues to increase [2]. Therefore, the adverse effects of noise on ventilation ductwork systems can offset the benefits of improving indoor air quality. Although the definition that noise is an undesirable and unpleasant sound has been widely accepted, it involves a sophisticated psychological process based on the perception of sound [3]. As a result, noise management [4] cannot simply pertain to sound pressure level (SPL) reduction. Ma et al. [3] found that the universal fundamental underlying structure of the human perception of sounds consisted of three latent perceptual dimensions: subjective evaluation of general judgement (Evaluation, E), sensitivity to magnitude (Potency, P), and sensation of temporal and spectral compositions of sound (Activity, A). These three perceptual dimensions, representing the factorial structure of human perception of sounds, are collectively referred to as the EPA model. Psychoacoustics is a branch of psychophysics that investigates the objective characteristics and subjective perceptual influence of noisy indoor environments. The psychoacoustic approach, which considers both auditory and non-auditory effects of noise, can provide an accurate prediction of human perceptual judgements of noise. With the support of objective psychoacoustic metrics and subjective psychometric properties, the multidimensional psychological impacts on the human perception of noise can be effectively evaluated and predicted.

## 5.2 OBJECTIVE PSYCHOACOUSTIC APPROACH

When sound waves propagate into the human cochlea, the basilar membrane of the organ of Corti is excited. Auditory signals are then transduced

DOI: 10.1201/9781003201168-5

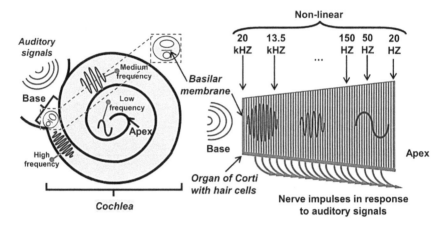

*Figure 5.1* Simplified schematic of basilar membrane in organ of Corti for characterising high-to-low frequency auditory signals from base to apex.

by hair cells, i.e., sensory cells in the organ of Corti, to electrical nerve impulses travelling to the brain. Because the resonance of sound waves of different frequencies is particularly intense at different regions of the basilar membrane, frequency dependence of human hearing depends on this (Figure 5.1). In 1933, Harvey Fletcher [5] introduced the concept of critical bands to describe the frequency bandwidth of 'band-pass auditory filters' created by particular regions of the basilar membrane. Subsequently, Eberhard Zwicker proposed the psychoacoustical scale, 'Bark scale', in 1961. A total of 24 critical bands in Bark scale were directly measured in a series of loudness approximation experiments [6]. Bark scale is one of the most popular psychoacoustic scales applied worldwide. This is because its complex band-pass filter can more accurately approximate the non-linear behaviour of the human auditory system than the 1/3-octave band-pass filter used in the traditional acoustic approach on a logarithmic scale (Figure 5.2). The numerical details of the two scales are shown in Figure 1.1 in Chapter 1.

## 5.2.1 Total loudness (*N*)

The traditional norm for assessing indoor environments is based on noise indices, such as noise criteria (*NC*), nose rating (*NR*), room constant ($R_C$), and equivalent continuous SPL ($L_{Aeq}$) [7, 8]. Although they can provide a single-value assessment of the acceptability of indoor noise environments, the noise features of environments and their psychological impact

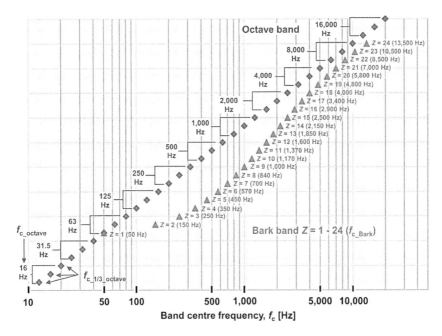

*Figure 5.2* Plot of centre frequency of octave (1/3 octave) bands and Bark bands in logarithmic scale.

on occupants (other than occupant sensitivity to noise magnitude) have not been well assessed [9–11]. Psychoacoustic metrics are used to estimate actual sensations of sounds based on human auditory responses to both energy and spectral contents of sound [12]. In assessing the acoustic comfort of indoor environments, the performance of the psychoacoustic metric $N$ is better than those of $NC$ and $NR$ [13].

Before calculating $N$, the specific loudness ($N'$) of each critical band (Bark band) must be calculated. According to Stevens' law of psychophysics [14], the intensity of sound wave stimulus (in terms of excitation) is hypothesised to be proportional to the perceived magnitude of the loudness sensation:

$$N'(E) \propto E^k, \tag{5.1}$$

where $E$ is the excitation level of a stimulus entering the organ of Corti; $N'$ is the specific loudness of loudness sensation at a critical band; and $k$ is the exponential constant of loudness sensation. When a noticeable change

($\Delta N'$) occurs due to a change in excitation ($\Delta E$), Eq. 5.1 can be expressed as Eq. 5.2:

$$\frac{\Delta N'}{N' + N'_{MT}} = k\frac{\Delta E}{E + E_{MT}}, \qquad (5.2)$$

where $N'_{MT}$ and $E_{MT}$ are the specific loudness and reference excitation of the masked threshold, respectively. The masked threshold is the internal noise floor, which is considerably less than the values of $N'$ and $E'$. The relationship between the masked threshold and threshold in quiet is described by the threshold factor ($s$) given by Eq. 5.3:

$$E_{MT} = \frac{E_T}{s}, \qquad (5.3)$$

where $E_T$ is the excitation of the threshold in quiet and $E_{MT}$ is the excitation of masked threshold produced by a test tone and within the critical band at the test-tone frequency.

By considering Eq. 5.2 as a differential equation, it can be solved by integrating both sides. The solution of this equation is as follows:

$$\mathrm{Ln}\left(N' + N'_{MT}\right) = k\ln\left(E + E_{MT}\right) + C, \qquad (5.4)$$

where $C$ is an unknown constant of the integration. Using reasonable boundary conditions, i.e., $N = 0$ and $E = 0$, the constant $C$ is found. Substituting $C = \ln\left(N'_{MT}\right) - k\ln\left(E_{MT}\right)$ into Eq. 5.4 yields Eq. 5.5, as follows:

$$\ln\left(\frac{N'}{N'_{MT}} + 1\right) = k\ln\left(\frac{E}{E_{MT}} + 1\right)$$

or

$$N' = N'_{MT}\left[\left(\frac{E}{E_{MT}} + 1\right)^k - 1\right]. \qquad (5.5)$$

After substituting Eq. 5.3 into Eq. 5.5, Eq. 5.6 is derived, as follows:

$$N' = N'_{MT}\left[\left(\frac{sE}{E_T} + 1\right)^{k_1} - 1\right]. \qquad (5.6)$$

By considering a reference specific loudness, $N'_{ref}$, of the reference excitation, $E_{ref}$, the following expression is derived:

$$\frac{N'_{MT}}{N'_{ref}} = \left(\frac{E_{MT}}{E_{ref}}\right)^k = \left(\frac{E_T}{sE_{ref}}\right)^k. \tag{5.7}$$

By substituting Eq. 5.7 into Eq. 5.6, the final expression is

$$N' = N'_{ref}\left(\frac{E_T}{sE_{ref}}\right)^k\left[\left(\frac{sE}{E_T}+1\right)^{k_1}-1\right]. \tag{5.8}$$

Using $E_{ref} = 10^{-12}$ W/m², $k_1 = 0.23$, and $s = 0.5$, the additional boundary condition corresponding to the N value of a 40-dB 1-kHz pure tone is equal to 1 sone. By replacing '1' inside the brackets in Eq. 5.8 by $(1 - s)$, the expression becomes

$$N' = 0.08\left(\frac{E_T}{E_{ref}}\right)^{0.23}\left[\left(0.5\frac{E}{E_T}+0.5\right)^{0.23}-1\right]\frac{sone}{Bark}. \tag{5.9}$$

Note that the additional boundary condition is approximately equal to N. The constant in Eq. 5.9 cannot can be obtained by simply inputting $N' = 1$; the proper approach must be

$$N_{40db,1-kHz} = \int N' d\text{Bark} = 1\,sone. \tag{5.10}$$

For any broadband sound, the total loudness is the sum of all $N'$ values of all critical banks (24 Barks):

$$N = \int_{Bark=1}^{24} N' d\text{Bark}\,(sone). \tag{5.11}$$

Details of the values of $k$ and $s$ are found in Eberhard Zwicker's book [12] and the computation guideline of N is found in ISO-532, which is an international standard [15].

## 5.2.2 Sharpness (S)

Although the spectral content of sounds is considered in the calculation of N, only the effect of this content on the loudness sensation is judged. The psychoacoustic metric $S$ is designed to characterise the skewness of energy distribution to high-frequency components. This is achieved by introducing an additional frequency weighting for the human sharpness sensation, as follows:

$$g(z) := \begin{cases} 1, & z \leq 14 \\ 0.00012z^4 - 0.0056z^3 + 0.1z^2 - 0.81z + 3.51, & z > 14 \end{cases}, \quad (5.12)$$

where $g(z)$ is critical-band-rate dependent (Figure 5.3).

The unit of $S$ is acum, which means 'sharp' in Latin; a 60-dB 1-kHz pure tone or narrow-band sound within the ninth Bark band is equal to 1 acum. The equation for $S$ is as follows:

$$S = 0.11 \frac{\int_0^{24\,Bark} N'g(z)z\,dz}{\int_0^{24\,Bark} N'\,dz} (acum). \quad (5.13)$$

In the foregoing equation, the denominator is the total loudness, N, implying that S is a normalised metric. The comparison of S values is a comparison

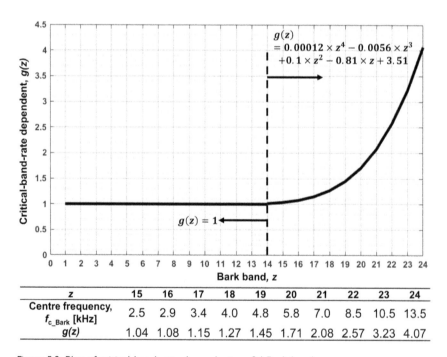

| z | 15 | 16 | 17 | 18 | 19 | 20 | 21 | 22 | 23 | 24 |
|---|---|---|---|---|---|---|---|---|---|---|
| Centre frequency, $f_{c\_Bark}$ [kHz] | 2.5 | 2.9 | 3.4 | 4.0 | 4.8 | 5.8 | 7.0 | 8.5 | 10.5 | 13.5 |
| g(z) | | 1.04 | 1.08 | 1.15 | 1.27 | 1.45 | 1.71 | 2.08 | 2.57 | 3.23 | 4.07 |

*Figure 5.3* Plot of critical-band-rate dependent on 24 Bark bands.

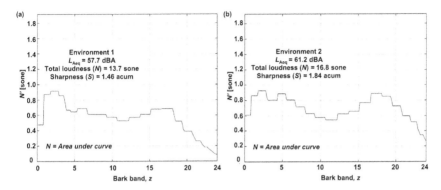

*Figure 5.4* Spectral envelope of (a) environment 1 and (b) environment 2 in terms of spectrum of specific loudness (*N'*) in 24 Bark bands.

of the skewness of the energy distribution (spectral envelope) of sounds to high-frequency components, as shown in Figure 5.4. A higher skewness of the energy distribution to high-frequency components is observed in environment 2, which has a higher *S* value.

Ma et al. [16] found that the *S* value is a valuable index for evaluating the health risks of people in an environment with high-frequency noise. The increase in low-frequency noise from ventilation systems can be one of the solutions to counteract the health risk from high-frequency noise; this can be achieved by reducing the *S* value in the environment. The results provide another perspective for noise control and show that SPL reduction is not the only means for managing noise.

## 5.2.3 Roughness (R)

In real-life situations, noise is a complex sound comprising different frequency components instead of a pure test tone of a single frequency. The human hearing system is unable to differentiate the components of a frequency pair that have the same amplitude and stimuli in the same critical band. Although more than one frequency component exists, the same region in the basilar membrane is excited. Otherwise, a feeling of roughness is created if the sound consists of multiple frequency components of different amplitudes and are in multiple critical bands. Frequency and temporal resolution are the two main factors considered in the calculation of the psychoacoustic metric, *R*:

$$R \sim f_{mod}\Delta L, \tag{5.14}$$

where $f_{mod}$ is the frequency for modulating the effect of frequency resolution (Hz), and $\Delta L$ is the temporal depth for masking the effect of temporal resolution (dB). For a more precise estimation of $R$, the following is proposed:

$$R \sim f_{mod} \int_0^{24 \, Bark} \Delta L(z) \, dz. \tag{5.15}$$

The unit of $R$ is asper, which means 'rough' in Latin; a 60-dB 1-kHz tone with 100% amplitude modulation at 70 Hz is equal to 1 asper. After introducing this boundary condition, the equation for $R$ becomes

$$R = 0.0003 f_{mod} \int_0^{24 \, Bark} \Delta L(z) \, dz \, (asper). \tag{5.16}$$

The experimental results of Zwicker et al. [12] showed that the roughness sensation of sounds was negligible if $f_{mod} < 15$ Hz or $f_{mod} > 300$ Hz.

### 5.2.4 Fluctuation strength (FS)

If the modulation frequency of sound is low ($f_{mod} < {\sim}20$), a sensation of fluctuating strength rather than a sensation of roughness is generated. Because the sensation of fluctuation is found to be strongest if $f_{mod} = 4$ Hz, the empirical equation considering the effects of frequency and temporal resolutions is as follows:

$$FS \sim \frac{\Delta L}{\left( \dfrac{f_{mod}}{4} + \dfrac{4}{f_{mod}} \right)}. \tag{5.17}$$

The unit of $FS$ is vacil, from the Latin 'vacilare'; a 60-dB 1-kHz tone with 100% amplitude modulation of 4 Hz is equal to 1 vacil. After introducing this boundary condition, the equation of $FS$ becomes

$$FS = 0.008 \frac{\int_0^{24 \, Bark} \Delta L(z) \, dz}{\left( \dfrac{f_{mod}}{4} + \dfrac{4}{f_{mod}} \right)} (vacil). \tag{5.18}$$

### 5.3 SUBJECTIVE PSYCHOACOUSTIC APPROACH

The objective psychoacoustic approach using psychoacoustic metrics and the subjective approach by measuring human responses to the acoustic environment are equally important in psychoacoustic studies. Subjective measurement data are crucial to determining how human perception is affected by the acoustic environment and the mechanism behind human perception. Therefore, the reliability and validity of the study's findings are critically challenged if inappropriate subjective measurement tools are used. For the objective psychoacoustic approach, the accuracy and precision of the measurement can be improved using better acoustic equipment with higher standards. Over the years, several investigators have endeavoured to devise a method for assessing the perceptual influence of environmental noise on occupants [17–21]. However, the heterogeneous nature of different psychoacoustic studies has increased the difficulty of standardising subjective assessment methods. Fortunately, this problem has been resolved by the critical systematic review and meta-analysis of Ma et al. [3]. Their findings provide insights into the fundamental mechanism of the human perception of sounds. Subsequently, a psychometric tool, the psychoacoustics perception scale (PPS), was designed to quantify human subjective responses. The PPS was found to be a valid, reliable, and applicable assessment tool for any built environment [22]. In the future, PPS is anticipated to be a valuable psychometric tool for standardising the subjective psychoacoustic approach of studies.

### 5.3.1 Rating methods

A semantic differential scale was applied to the PPS. This scale, proposed by Osgood in 1952, is a commonly used psychological scale [23] to represent the 'meaning of things' quantitatively. The first psychoacoustic approach using a semantic differential scale to represent the 'human perceptions of sounds' quantitatively was implemented by Solomon in 1959 [24]. A systematic review conducted by Ma et al. [3] integrated the significant results of all accessible English language psychoacoustic studies in finding 'human perceptions of sounds' after 1959. The meta-analysis of the results of eligible studies extracted universal findings. The three fundamental human perceptual dimensions of sounds are found to be 'Evaluation (E)', 'Potency (P)', and 'Activity (A)', pertaining to the human general judgement, sensitivity to magnitude, and sensation of the temporal and spectral compositions of perceived sound, respectively (Figure 5.5).

The PPS is a psychometric tool consisting of nine semantic differential questions. Each question is formed by a bipolar adjective pair (AP) with two descriptors of opposite meanings. The score for each item in the PPS ranges from –3 to +3. For example, the AP score of '*Pleasant–Unpleasant*' has seven levels: 'Extremely Pleasant (–3)', 'Quite Pleasant (–2)', 'Slightly Pleasant (–1)', 'Equally (0)', 'Slightly Unpleasant (1)', 'Quite Unpleasant (2)', and 'Extremely Unpleasant (3)'. The E, P, and A scores are defined as the sum of the three compulsory APs in the factors. The total EPA score is defined as the sum of all nine APs in the PPS. The 'human perceptions of

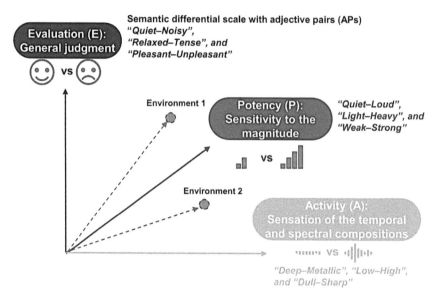

*Figure 5.5* Schematic of semantic spaces formed by three fundamental human perceptual dimensions of sounds: 'Evaluation', 'Potency', and 'Activity'.

sounds' can be represented by their positions in the semantic spaces of E, P, and A scores. The foregoing depicts a general picture of the perceived acoustic environment, facilitating the comparison of sounds in different aspects rather than merely comparing magnitudes.

## 5.3.2 Adjective pair (AP, Descriptors)

The proper selection of questions is the first and most important step in any subjective assessment. Improper selection by including redundant and irrelevant questions lengthens the time of assessment, increases the difficulty of subject recruitment, reduces the concentration of subjects on the assessment, and affects the reliability of measurement results [25]. In contrast, the lack of important questions limits the validity and generality of the assessment results. In a systematic review by Ma et al. [3], a total of 5 677 participants, 828 756 ratings, 1 365 sounds, and 686 descriptors were included in the 45 eligible studies on the assessment of human perceptions of various indoor and outdoor sounds using a semantic differential scale. Quantitative meta-analysis was based on the extracted data of the 27 APs applied to the eligible studies after an important analysis. All 27 APs were significant in assessing the human perception of sound. The meta-analysis of the extracted data further confirmed that a minimum of nine APs were required to represent the acoustic environment quantitatively in

*Table 5.1* Eighteen APs associated with three fundamental human perceptual dimensions of sounds

| Adjective pair | Compulsory item (perceptual dimension) | Associated with perceptual dimension(s) |
| --- | --- | --- |
| **Quiet –Noisy** | ✓ Evaluation (E) | E |
| **Relaxed–Tense** | ✓ Evaluation (E) | E |
| **Pleasant–Unpleasant** | ✓ Evaluation (E) | E |
| **Quiet–Loud** | ✓ Potency (P) | P |
| **Light–Heavy** | ✓ Potency (P) | P |
| **Weak–Strong** | ✓ Potency (P) | P |
| **Deep–Metallic** | ✓ Activity (A) | A |
| **Low– High** | ✓ Activity (A) | A |
| **Dull–Sharp** | ✓ Activity (A) | A |
| **Warm–Cold** | O | E |
| **Beautiful–Ugly** | O | E |
| **Comfortable–Uncomfortable** | O | E and P |
| **Like–Dislike** | O | E and A |
| **Calming–Agitating** | O | E and A |
| **Harmonic–Discordant** | O | E and A |
| **Bright–Dark** | O | E and A |
| **Soft–Hard** | O | E, P, and A |
| **Gentle–Violent** | O | E, P, and A |

three fundamental human perceptual dimensions of sounds. The three most suitable APs for assessing the general human judgement of sound in the E dimension are '*Quiet–Noisy*', '*Relaxed–Tense*', and '*Pleasant–Unpleasant*'. The three most suitable APs for assessing the human sensitivity to the magnitude of sound in the P dimension are '*Quiet–Loud*', '*Light–Heavy*', and '*Weak–Strong*'. The three most suitable APs for assessing the human sensation of the temporal and spectral compositions of sound in the A dimension are '*Deep–Metallic*', '*Low–High*', and '*Dull–Sharp*'. The meta-analysis results also show a potential association between the other perceptions and the E, P, and A dimensions (Table 5.1). The PPS only requires the inclusion of the nine compulsory APs in the E, P, and A dimensions; researchers have the freedom to assess other perceptions based on their own research purpose or interest. The use of PPS can aid in reducing the diversity in the selection of APs and hence also reduce the discrepancy among the analytical results of psychoacoustic studies.

## 5.4 VALIDITY AND RELIABILITY OF PPS

The PPS is the first designed psychometric tool based on the universal underlying structure of the human perceptual dimensions of sound [3]. The PPS utilizes the EPA model, which serves as an integrated representation of the three fundamental perceptual dimensions and their underlying factorial structure (Figure 5.6). It considerably simplifies the subjective assessment without missing any important human perception of sound. The PPS was further validated by Ma et al. as a reliable and applicable psychometric tool for qualitatively assessing the perceptions of the general judgement, energy content, and temporal and spectral contents of sound [22]. The high values of the test–retest Cronbach's $\alpha$s and composite reliabilities of compulsory items in the E, P, and A dimensions demonstrate the satisfactory internal consistency of the items. The results also verified the reliability of the PPS in assessing the E, P, and A dimensions. They also supported the use of E, P, and A scores to represent the sound quality of the acoustic environment in the designed dimensions quantitatively. The E, P, and A scores can be treated as indices of the sound quality of the environment and used in analysing the environmental influence on occupants. In general, the factorial structure in terms of factor loading, scalar, factor covariance, and error variances of the P and A dimensions was found to be invariant across gender. The total EPA score can represent the joint contribution of the E, P, and A dimensions and serve as a significant predictor of other perceptions. The development of PPS may be valuable for psychoacoustic studies. The PPS is a potentially fundamental psychometric tool to standardise subjective assessment methods in the psychoacoustic approach.

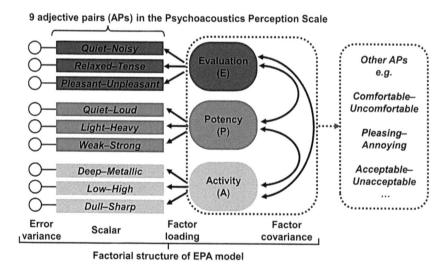

Figure 5.6 Schematic of factorial structure behind the PPS.

## 5.5 PSYCHOACOUSTIC PREDICTION METHOD

### 5.5.1 Objective psychoacoustic metric

The prediction of psychoacoustic metrics [26, 27] can be improved by considering the effects of spatial variation (Figure 5.7). Improved environmental noise control [28] can be achieved by accurately predicting the noise of building elements [29–33] and measuring the noise outdoors [34–36] and indoors [37]. Sound attenuation during propagation is not a new concept in acoustics. The inverse square distance law for the distance dependence of sound intensity is well known. Ma et al. [26] found that the effects of distance should be considered using sound energy content-related metrics. Moreover, the psychoacoustic metric, $N$, of the perceived loudness sensation of subjects was *predictable* in sound propagation. A subjective life experience in which loudness decreased as the distance increased was found. Their study is the first to provide a scientific and objective method for predicting the decrement in loudness using psychoacoustic metrics. The study results [26] provide a new approach to estimate the values of $N$ at different positions successfully from a noise source. The numerical approximation of $N$ in the inverse power, 0.46 of $D$, was firstly proposed based on the study results. Owing to the non-linearity of logarithmic functions in the unweighted SPL $(L_Z)$ and A-weighted SPL $(L_A)$ calculations, the magnitude of environmental change cannot be easily interpreted by the changes in $L_Z$ and $L_A$ values.

For example, the magnitudes of changes in environmental sound energy and occupants' loudness perception of a 6-dBA decrement in 30-dBA and 50-dBA rooms are different. The changes in $L_Z$ and $L_A$ values may not be suitable metrics for the statistical analysis and prediction of environmental effects on subjective responses. The $N$ measurement serves as an alternative to environmental assessment and prediction. Because the unit of $N$ is a linear scale (sone scale), the difference among $N$ values can be interpreted. As mentioned, the resulting 6-dBA decrement when the distance from the sound source is doubled cannot clearly explain the environmental influence on acoustic properties or subjective responses. In contrast, the decrease in $N$ value during sound propagation allows people to perceive the extent to which the environment has quieted down. The offhand prediction of psychoacoustic metrics is now possible. In Eq. 5.9, when the excitation (sound intensity) of sound considerably exceeds that of $E_{ref}$ ($10^{12}$ W·m$^{-2}$), $N'$ is approximated to a power 0.23 of the excitation. Therefore, $N'$ was approximated to the inverse power of 0.46 of the distance ($D$) from the noise source. In other words, doubling the distance from the sound source causes the value of $N$ to be 0.73 times smaller (Eq. 5.19). The predictability of psychoacoustic metrics with spatial variation is demonstrated for the first time:

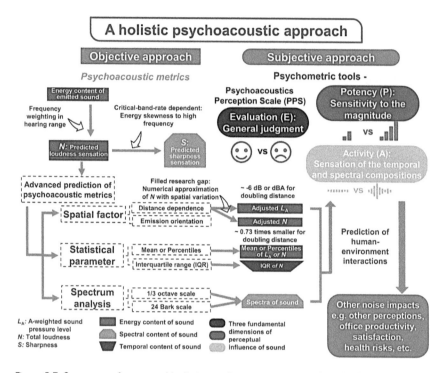

*Figure 5.7* Summary of proposed holistic psychoacoustic approach to building acoustics.

$$\frac{N_{D'}}{N_D} \sim \left(\frac{D'}{D}\right)^{-0.46} , \qquad (5.19)$$

where $N_D$ ($N_{D'}$) is the total loudness at distance $D$ ($D'$) from the noise source. The effect of emission orientation was observed when the distance from the sound source was less than 0.3 m. The properties of sound energy content were characterised by metrics $L_A$ and $N$, whereas those of the sound spectral content were characterised by metric $S$. In contrast, the spectral content-related properties were not sensitive to changes in spatial variation. The findings fill the research gap in psychoacoustic measurements and provide guidance for the future environmental assessment and prediction of sound energy content as well as temporal and spectral contents.

## 5.5.2 Subjective response

The PPS development provides a guideline applicable to the construction of a subjective scale for assessing the built environment. Moreover, it promotes the possibility of psychoacoustic prediction from objective acoustic properties to subjective responses. Occupant-oriented decision-making can be achieved in future building designs by providing acceptable psychoacoustic prediction. A holistic psychoacoustic approach can also aid in understanding the human–environment interactions between the acoustic environment, human perceptions, and other potential noise impacts, such as office productivity [38], satisfaction [39], and health risks [16, 40]. The foregoing is key to sustainable noise control designs for building ventilation systems [41]. The study conducted by Ma et al. showed that the monitoring of $S$ and $N$ of noise was required because it could serve as an indicator for assessing noise impacts on indoor environments [16]. Although similar results were found between the measurements of $L_{Aeq}$ and $N$ of the overall perceived loudness of noise, the $N'$ spectral analysis demonstrated its ability to assess the corresponding loudness changes from different frequency components of noise. The capacity to identify perceived loudness changes from different noise sources, such as ventilation systems, is important for environmental noise evaluation, especially for indoor environments. Further attention must also be given to the frequency components of indoor environments. This is because noise sharpness is associated with negative short-term psychological symptoms affecting the states of hearing and health in occupations. The calculation of $S$ provided more specific information on the sound quality of noise than on overall loudness. Thus, the quantification of the sound quality of noise aids in understanding the effects of noise on the health condition related to occupation. The results of the study showed the capacity of psychoacoustics metrics in quantifying the sound quality of noise and estimating its negative impact on health associated with occupation. The energy and spectral contents of the indoor environment affect the prevalence of long-term and short-term negative health symptoms, respectively [16].

In the studies of Ma et al. [42, 43], the temporal content of the environment in terms of the statistical noise level of psychoacoustic metrics (e.g., $N_{10}$, $N_{50}$, and $N_{90}$) can aid in predicting environmental noise impacts. In general, the acoustic influence of a significant noise source can be represented by $L_{A10}$, $N_{10}$, and $S_{10}$, and ambient noise can be represented by $L_{A50}$, $N_{50}$, $L_{Aeq}$, $N_{eq}$, and $S_{eq}$. The statistical analysis of $N$ disclosed the perceptual influence on the noise sensitivity of occupants, further affecting their behaviour. The statistical analysis of $S$ also reveals that this metric is a significant indicator of the behavioural influence of noise on occupants. For long-term development, the impact of metrics $L_{A90}$, $N_{90}$, and $S_{90}$ of the background noise level on general occupational health was also determined. These findings provide insights into future noise management to assess, monitor, evaluate, predict, and control the effect of noise on occupants [43].

In dealing with maximum noise levels, metrics $L_{Amax}$, $N_{max}$, and $L_{A10}$–$L_{A90}$ can aid in distinguishing the variations in the perceptions of subjects between two different environments [42]. The perception of sound variation was more related to the variation in spectral content than to the energy content of sounds. Traditional $L_{Aeq}$ monitoring is insufficient for predicting the perceptual influence on people. Monitoring the maximum noise level and other metrics regarding the spectral and temporal contents of the acoustic environment is recommended. The analyses of preferences can provide insight into the balance between the space utilisation of the built environment and environmental improvement. High, medium, and low preferences were found for the natural, human-made, and mechanical sounds in the environment, respectively [42]. The insertion of natural sounds and the elimination of mechanical sounds are the two main directions of acoustic improvement. The responses of subjects can provide important information to the further design of ventilation noise systems.

## 5.6 SUMMARY

This chapter introduces fundamental knowledge on traditional objective psychoacoustic metrics and up-to-date subjective psychometric (PPS) in the psychoacoustic approach. Examples of holistic psychoacoustic assessments and predictions of ventilation noise are discussed in Chapter 6.

## References

1. World Health Organization. Regional Office for Europe, 2011 *Burden of disease from environmental noise: Quantification of healthy life years lost in Europe.*
2. W. Passchier-Vermeer and W.F. Passchier, 2000 *Environmental Health Perspectives* 108, pp.123–131, Noise exposure and public health.
3. K.W. Ma, H.M. Wong, and C.M. Mak, 2018 *Building and Environment* 133, pp.123–150, A systematic review of human perceptual dimensions of sound: Meta-analysis of semantic differential method applications to indoor and outdoor sounds.
4. B. Berglund, T. Lindvall, D.H. Schwela, and World Health Organization, 1999, Guidelines for community noise.
5. H. Fletcher and W.A. Munson, 1933 *Bell System Technical Journal* 12, pp.377–430, Loudness, its definition, measurement and calculation.
6. E. Zwicker, 1961 *The Journal of the Acoustical Society of America* 33, pp.248–248, Subdivision of the audible frequency range into critical bands (Frequenzgruppen).
7. International Organization for Standardization, *Acoustics – Description, measurement and assessment of environmental noise – Part 1: Basic quantities and assessment procedures*, 2016.
8. Acoustical Society of America, *Criteria For Evaluating Room Noise*, 2019.
9. U. Ayr, E. Cirillo, and F. Martellotta, 2001 *Applied Acoustics* 62, pp.633–643, An experimental study on noise indices in air conditioned offices.
10. B. Hay and M.F. Kemp, 1972 *Journal of Sound and Vibration* 23, pp.363–373, Measurements of noise in air conditioned, landscaped offices.

11. S.K. Tang and M.Y. Wong, 2004 *Journal of Sound and Vibration* 274, pp.1–12, On noise indices for domestic air conditioners.

12. E. Zwicker and H. Fastl, 2003 *Psychoacoustics: Facts and models.* Springer Science & Business Media: Berlin/Heidelberg.

13. S.K. Tang, 1997 *The Journal of the Acoustical Society of America* 102, pp.1657–1663, Performance of noise indices in air-conditioned landscaped office buildings.

14. S.S. Stevens, 1957 *Psychological Review* 64, p.153, On the psychophysical law.

15. International Organization for Standardization, 2017 *ISO 532-1: Acoustics–Methods for calculating loudness–Part 1: Zwicker method.* International Organization for Standardization: Geneva, Switzerland.

16. K.W. Ma, H.M. Wong, and C.M. Mak, 2017 *International Journal of Environmental Research and Public Health* 14, p.1084, Dental environmental noise evaluation and health risk model construction to dental professionals.

17. E. Bjork, 1985 *Acustica* 58, pp.185–188, The perceived quality of natural sounds.

18. M. Galiana, C. Llinares, and Á. Page, 2012 *Building and Environment* 58, pp.1–13, Subjective evaluation of music hall acoustics: Response of expert and non-expert users.

19. J.G. Ih, S.W. Jang, C.H. Jeong, and Y.Y. Jeung, 2009 *Journal of Vibration and Acoustics, Transactions of the ASME* 131, pp.0345021–0345025, A study on the sound quality evaluation model of mechanical air-cleaners.

20. J.Y. Jeon, P.J. Lee, J.Y. Hong, and D. Cabrera, 2011 *The Journal of the Acoustical Society of America* 130, p.3761, Non-auditory factors affecting urban soundscape evaluation.

21. J.Y. Jeon, J. You, C.I. Jeong, S.Y. Kim, and M.J. Jho, 2011 *Building and Environment* 46, pp.739–746, Varying the spectral envelope of air-conditioning sounds to enhance indoor acoustic comfort.

22. K.W. Ma, C.M. Mak, and H.M. Wong, 2020 *Journal of Building Engineering* 29, p.101177, Development of a subjective scale for sound quality assessments in building acoustics.

23. C.E. Osgood, 1952 *Psychological Bulletin* 49, p.197, The nature and measurement of meaning.

24. L.N. Solomon, 1959 *The Journal of the Acoustical Society of America* 31, pp.492–497, Search for physical correlates to psychological dimensions of sounds.

25. J.A. Gliem and R.R. Gliem, 2003, Calculating, interpreting, and reporting Cronbach's alpha reliability coefficient for Likert-type scales. *Midwest research-to-Practice Conference in Adult, Continuing, and community education.* Ohio State University: Columbus, Ohio.

26. K.W. Ma, C.M. Mak, and H.M. Wong, 2020 *Applied Acoustics* 168, p. 107450, Acoustical measurements and prediction of psychoacoustic metrics with spatial variation.

27. C.M. Mak and Z. Wang, 2015 *Building and Environment* 91, pp.118–126, Recent advances in building acoustics: An overview of prediction methods and their applications.

28. C.Z. Cai and C.M. Mak, 2016 *Journal of the Acoustical Society of America* 140, pp.EL471–EL477, Noise control zone for a periodic ducted Helmholtz resonator system.

29. C.M. Mak and W.M. Au, 2009 *Applied Acoustics* 70, pp.11–20, A turbulence-based prediction technique for flow-generated noise produced by in-duct elements in a ventilation system.

30. C.M. Mak, 2005 *Journal of Sound and Vibration* 287, pp.395–403, A prediction method for aerodynamic sound produced by multiple elements in air ducts.

31. C.M. Mak and J. Yang, 2000 *Journal of Sound and Vibration* 3, pp.743–753, A prediction method for aerodynamic sound produced by closely spaced elements in air ducts.

32. C.M. Mak, J. Wu, C. Ye, and J. Yang, 2009 *Journal of the Acoustical Society of America* 125, pp.3756–3765, Flow noise from spoilers in ducts.

33. C.M. Mak, 2002 *Applied Acoustics* 63, pp.81–93, Development of a prediction method for flow-generated noise produced by duct elements in ventilation systems.

34. C. Cai, C.M. Mak, and X. He, 2019 *Noise & Health* 21, p.142, Analysis of urban road traffic noise exposure of residential buildings in Hong Kong over the past decade.

35. C.M. Mak, W. Leung, and G. Jiang, 2010 *Building Services Engineering Research and Technology* 31, pp.131–139, Measurement and prediction of road traffic noise at different building floor levels in Hong Kong.

36. C.M. Mak and W.S. Leung, 2013 *Environmental Engineering & Management Journal (EEMJ)* 12, pp. 449–456, Traffic noise measurement and prediction of the barrier effect on traffic noise at different building levels.

37. W.M. To, C.M. Mak, and W.L. Chung, 2015 *Noise Health* 17, p.429, Are the noise levels acceptable in a built environment like Hong Kong?

38. C.M. Mak and Y. Lui, 2012 *Building Services Engineering Research and Technology* 33, pp.339–345, The effect of sound on office productivity.

39. P. Xue, C.M. Mak, and Z.T. Ai, 2016 *Energy and Buildings* 116, pp.181–189, A structured approach to overall environmental satisfaction in high-rise residential buildings.

40. Z.T. Ai, C.M. Mak, and H.M. Wong, 2017 *Building Services Engineering Research and Technology* 38, pp.522–535, Noise level and its influences on dental professionals in a dental hospital in Hong Kong.

41. C.M. Mak, W. To, T. Tai, and Y. Yun, 2015 *Indoor and Built Environment* 24, pp.128–137, Sustainable noise control system design for building ventilation systems.

42. K.W. Ma, C.M. Mak, and H.M. Wong, 2021 *Applied Acoustics* 171, p.107570, Effects of environmental sound quality on soundscape preference in a public urban space.

43. K.W. Ma, C.M. Mak, and H.M. Wong, 2020 *Applied Acoustics* 161, p.107164, The perceptual and behavioral influence on dental professionals from the noise in their workplace.

# Chapter 6

# Case studies and examples

## 6.1 INTRODUCTION

Case studies or examples are presented in the following sections to illustrate the prediction of transmission noise, flow-generated noise, vibration isolation performance, and human psychoacoustic responses discussed in Chapters 2, 3, 4, and 5, respectively.

## 6.2 CALCULATION OF NOISE TRANSMISSION IN AIR DUCT

The following is a typical example of the acoustic design of a ventilation ductwork system. Flow-generated noise and break-out noise are ignored in this calculation. An open-plan office with dimensions of 10 m × 8 m × 3 m is provided with a ventilation ductwork system, as shown in Figure 6.1. For noise control, a dissipative silencer is installed in the air distribution ductwork system. The reverberation time in the office is assumed to be 0.9 s. The total amount of conditioned air is evenly distributed from an air-handling unit to the office through six 0.1 m² supply air outlets. The air ducts are plain without any internal acoustic lining. The critical point of noise in the room is assumed to be 1.2 m directly under any supply air outlet. The design criterion for the maximum allowable noise level, $L_{p, Allow}$, in the room is assumed to be 35 dB at the 2-kHz octave-band centre frequency; the sound power level of the fan at this frequency is 82 dB. The length of the 450 × 400-mm main air duct from the silencer to the first branch is 8 m, and the length of the 350 mm × 250 mm branch air duct is 1.8 m. To estimate the required attenuation given by the dissipative silencer at 2 kHz, determining the total sound pressure at a point by logarithmic addition of direct and reverberant sound levels is necessary.

The following data are used to calculate the noise transmission:

- The attenuation per metre run of the 450 × 400-mm section of the main air duct at 2 kHz is assumed to be 0.125 dB/m.

DOI: 10.1201/9781003201168-6

- The attenuation of the 350 × 250-mm branch air duct at 2 kHz is insignificant.
- Attenuation due to other possible in-duct components can be ignored.
- The correction for the direct sound pressure level in the room corresponding to the outlet size and position at 2 kHz is assumed to be 9 dB.
- The average absorption coefficient in the office is less than 0.1.
- The end reflection from the 0.1m² outlet at high frequencies is assumed to be 0.

The estimation of the required attenuation, $L_{p,Atten}$, by the silencer at 2 kHz is as follows:

1. Determine the sound power level, $L_w$, at outlets:

   $L_w = 82 - 0.125 \times 8 = 81$ dB.

2. Determine the direct sound pressure level, $L_{p,Dir}$, at the critical point:

   $$L_{p,Dir} = 81 - 10\log 6 + 9 - 10\log\left(4\pi \times 1.2^2\right)$$

   $$= 81 - 7.8 + 9 - 12.6 = 69.6\,\text{dB}.$$

3. Determine the room constant, $R_C$, by first using the Sabine formula, $RT = \dfrac{0.161V}{S\bar{\alpha}}$, to obtain the mean absorption coefficient, $\bar{\alpha}$:

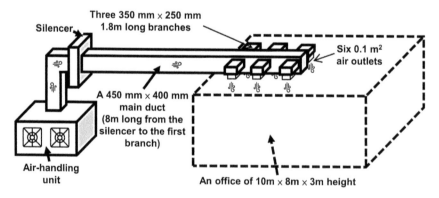

Figure 6.1  Diagram of air ductwork layout from air-handling unit to office.

$$\bar{\alpha} = \frac{0.161 \times V}{RT \times S} = \frac{0.161 \times (10 \times 8 \times 3)}{0.9 \times 2 \times (10 \times 8 + 10 \times 3 + 8 \times 3)}$$

$$= \frac{0.161 \times 240}{0.9 \times 268} = 0.16.$$

Then, compute $R_C$. For low absorption,

$$R_c = \frac{S\bar{\alpha}}{1 - \bar{\alpha}} = \frac{268 \times 0.16}{1 - 0.16} = 51.1.$$

4.  Determine the reverberant sound pressure level, $L_{p,Rev}$, from transmission noise:

$$L_{p,Rev} = L_w + 10\log\left(\frac{4}{R_c}\right) = 81 + 10\log\left(\frac{4}{51.1}\right) = 69.9 \text{ dB.}$$

5.  Determine the sound pressure level of total noise, $L_{p,Total}$, in the room:

$$L_{p,Total} = 10\log\left(10^{\frac{L_{p,Dir}}{10}} + 10^{\frac{L_{p,Rev}}{10}}\right)$$

$$= 10\log\left(10^{\frac{69.6}{10}} + 10^{\frac{69.9}{10}}\right) = 72.8 \text{ dB.}$$

6.  Determine the required attenuation, $L_{p,Atten}$, of silencer at 2 kHz:

$$L_{p,Atten} = L_{p,Total} - L_{p,Allow} = 72.8 - 35 = 37.8 \text{ dB.}$$

## 6.3 PREDICTION OF FLOW-GENERATED NOISE USING PRESSURE-BASED PREDICTION METHOD

### 6.3.1 Flow-generated noise experiment

An experiment was conducted by Mak et al. [1] to verify the results predicted by the two predictive equations for flow-generated noise (Eqs. 3.22 and 3.23). A schematic of the experimental test rig is shown in Figure 6.2(a). A three-phase variable-speed centrifugal flow fan constrained from vibration by four springs and placed in a lined acoustic enclosure was used to provide airflow. Silencers and acoustically lined elbows were installed to attenuate noise from the upstream and

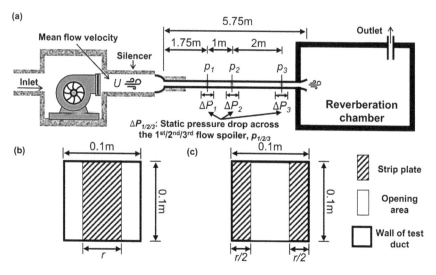

Figure 6.2 (a) Schematic of experimental test rig using two main types of spoilers; (b) centrally placed strip spoiler; and (c) two strip plates placed symmetrically at both sides of the duct leaving central vertical strip of duct open.

downstream sides. To reduce the breakout noise from the 0.1 × 0.1-m steel test air duct, a 25-mm-thick lining made of absorbing material was placed on each side of the duct. Hence, a quiet and fully developed flow through the first test piece starting from the inlet of airflow was obtained. The first flow spoiler was located at approximately 1.75 m away from the duct entrance section. The second and third spoilers were situated 1 and 3 m downstream from the first spoiler, respectively. The total length of the duct including the extended part of the reverberation chamber was 5.45 m. The chamber had an outlet cone of 0.16 m × 0.16 m and a length of 0.3 m for acoustic measurement. The chamber was also equipped with lined outlet ducts enabling air to escape without noise. The experimental test rig was located in another reverberation chamber to ensure that the level of sound measured in the system consistently exceeded the background noise level in the laboratory.

The spoilers used in the experiment were made of a 1-mm-thick steel plate to provide a relatively rigid obstruction without significant vibration in the air stream. The spoiler plates were fixed using springs and force transducers between the flanges of the two adjoining sections of the test duct. A compressed foam rubber was used to seal the gaps. Figure 6.2(b) and (c) shows the cross-section of the duct with the two main types of spoilers used in the experiment. A strip spoiler is centrally placed, and two strip

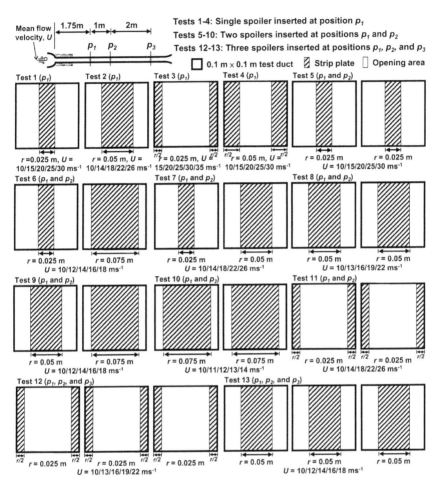

*Figure 6.3* Different spoiler configurations (Tests 1–13) under different mean flow velocities in experimental test rig shown in Figure 6.1.

plates are symmetrically placed on both sides of the duct by leaving a vertical opening at the centre of the duct. The height of the strip is equal to that of the test duct (0.1 m). The various spoiler widths used in the experiments are shown in Figure 6.2. Various spoiler configurations under different mean flow velocities are chosen to validate the prediction of flow-generated noise, as illustrated in Figure 6.3. The first four experiments (Tests 1, 2, 3, and 4) for each single spoiler are conducted to obtain the normalised spectrum, as shown in Figure 6.5. Seven experiments (Tests 5, 6, 7, 8, 9, 10, and

11) included two spoilers, whereas the last two experiments (Tests 12 and 13) considered three spoilers.

## 6.3.2 Normalisation of experimental results for single-flow spoiler

The measured in-duct sound power levels ($L_{w,D}$) for Test 1 in the experiment of Mak et al. [1] are shown in Figure 6.4, depicting the typical characteristics of flow-generated noise. For instance, peak $L_{w,D}$ occur at extremely low frequencies, and a systematic decrease in $L_{w,D}$ with increasing frequency is observed. In addition, the increase in $L_{w,D}$ with velocity is greater at frequencies exceeding the duct cut-on frequency. The measured values of the drag coefficient ($C_D$) and the $L_{w,D}$ of the experiment conducted by Mak et al. were normalised for Tests 1–4 based on the equations of Nelson and Morfey (Eqs. 3.14 and 3.15). They found that the collapse of experimental data is remarkable, as shown in Figure 6.5 [1]. The normalised results of all the tested spoilers agreed well with the trend lines of Nelson and Morfey [2]. A trend line based on simple linear relationships over the range of measurements was derived.

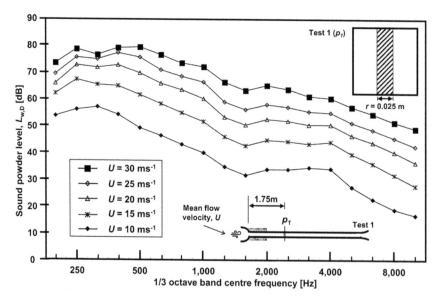

*Figure 6.4* 1/3-octave-band spectra of in-duct sound power level $L_{w,D}$ of flat-plate spoiler inserted at position $p_1$ and five different mean flow velocities in Test 1.

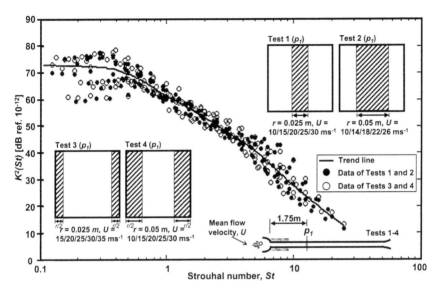

*Figure 6.5* Overall collapse of normalised data for all tested single spoilers in Tests 1–4.

### 6.3.3 Comparison between predicted and measured results

The normalised spectrum is shown in Figure 6.5. This figure and the coherence functions (i.e., two predictive equations for flow-generated noise (Eqs. 3.22 and 3.23)) for multiple in-duct flow spoilers are used to predict the $L_{w,D}$ produced by the two or three flow spoilers in Tests 5–13. The predicted and measured results of the $L_{w,D}$ in the nine tests are shown in Figures 6.6 and 6.7. Excellent and consistent agreement between the measured and predicted results for all test combinations of the spoilers was observed at most frequencies and mean duct flow velocities. The error between the predicted and measured results at most frequencies was ±0–3 dB except for some particular frequencies and mean flow velocity. The deviation between the predicted and measured results at certain frequencies at or exceeding 4 kHz for a particular flow may have been due to the vibration modes of the flow spoilers or the duct system at high frequencies.

The results suggest that the predictive equations for multiple in-duct flow spoilers are useful for predicting the level and spectral contents of the flow-generated noise produced by multiple in-duct flow spoilers at the design stage. This is because the prediction is based on a normalised spectrum,

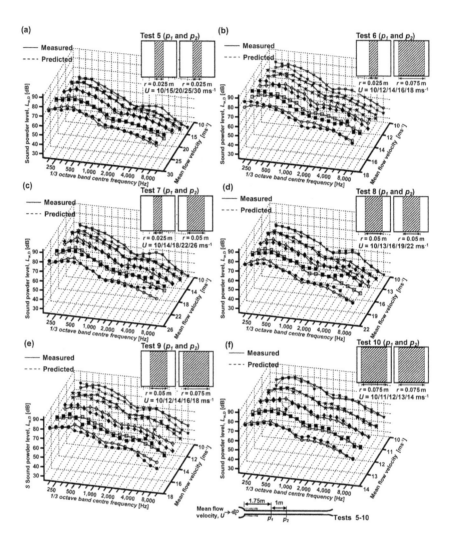

*Figure 6.6* Comparison between measured and predicted values of sound power level in Tests 5–10 ((a)–(f)) with two centrally placed strip spoilers inserted at positions $p_1$ and $p_2$ at different mean flow velocities.

and all the parameters used in the prediction can be obtained through measurements. Therefore, the predictive equations become the basis of the generalised prediction method for the flow-generated noise produced by multiple in-duct elements.

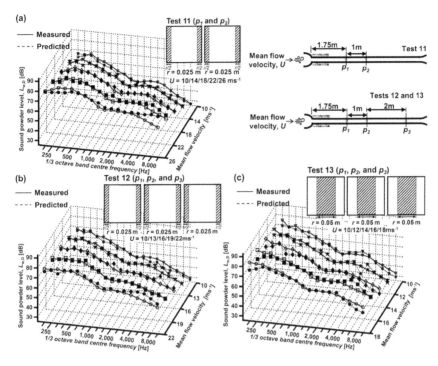

*Figure 6.7* Comparison between measured and predicted values of sound power level in (a) Test 11 with two spoilers containing two strip plates placed symmetrically at both sides of duct and inserted at positions $p_1$ and $p_2$ and in (b) Test 12 and (c) Test 13 with three different types of spoilers inserted at positions $p_1$, $p_2$, and $p_3$.

## 6.4 PERFORMANCE ASSESSMENT OF VIBRATION ISOLATION USING POWER TRANSMISSIBILITY METHOD

The work of Mak and Su considered a vibratory machine model with four symmetrical supports on a square concrete plate [2]. The model, shown in Figure 6.8, has the following physical parameters for the concrete plate: density, $\rho = 2.8 \times 10^3$ kg/m³; Young's modulus, $E = 2.1 \times 10^{10}$ N·m⁻²; loss factor, $\eta = 2 \times 10^{-2}$; and Poisson's ratio, $\mu = 0.2$. The geometrical dimensions are length ($l$) = 3.5 m, width ($W$) = 3.5 m, and the thickness is d = 0.24 m. The real and imaginary parts of the effective mobility of the contact point of the square concrete plate can be obtained from [3], as shown in Figure 6.9(a). The mobility value is extremely small at most frequencies but becomes relatively large when the frequency range is 48–52 Hz. The

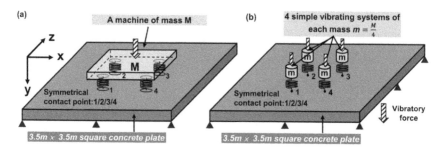

Figure 6.8 Schematics of (a) machine of mass $M$ with four symmetrical contact points and (b) four simple vibrating systems of each mass, $m$ ($m = M/4$), with four symmetrical contact points on a simply-simply-simply-simply square concrete plate.

power transmissibility, $\gamma$ (Eqs. 4.23 and 4.24 for single and four contact points, respectively), and dynamic forces transmitted to the floor against distributing frequencies if the mass, $m = 500$ kg (i.e., the machine mass, $M = 2,000$ kg) and the chosen natural frequency, $f_n = 15$ Hz are shown in Figure 6.9(b) and (c), respectively. As shown in Figure 6.9(c), the source-inherent vibratory force, $F_o$ ($F_o = j\omega M V_o$), is equal to 1.

Figure 6.9(b) shows that $\gamma$ is less than 1 at most frequencies (except 48–52 Hz); then, it decreases with increasing frequencies starting from 52 Hz. A peak $\gamma$ value (approximately 10) is observed when the frequency is 48–52 Hz. This indicates that the power transmitted to the floor increases when the spring isolator is installed at frequencies between 48 and 52 Hz because of the effect of large floor mobility.

Based on the definition of force transmissibility (Eq. 4.17), the force transmitted to the floor with a spring isolator via the contact point must be smaller than that without the spring isolator when $\omega/\omega_n > \sqrt{2}$. Figure 6.9(c) shows that at most frequencies (except for frequencies between 48 and 52 Hz), the force transmitted to the floor with the spring isolator through a contact point is considerably less than that without the spring isolator. The dynamic force transmitted to the floor with the spring isolator at frequencies between 48 and 52 Hz exceeded that transmitted to the floor without the spring isolator. Similarly, Figure 6.9(b) shows that the power at most frequencies (except for frequencies between 48 and 52 Hz) is reduced after the installation of the spring isolator ($\gamma < 1$). Therefore, at most frequencies, both the total power and dynamic force transmitted to the floor are reduced after the installation of the spring isolator. However, they increase at frequencies between 48 and 52 Hz, approximating the resonant frequency of the floor shown in Figure 6.9(a).

A vibratory machine with four symmetrical supports placed asymmetrically on the same plate, as shown in Figure 6.10, is considered. The same physical and geometric dimensions are used. Because the real and imaginary

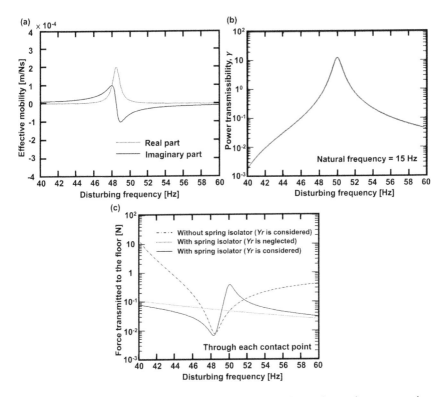

*Figure 6.9* Plots of (a) effective mobility of the simply-simply-simply-simply concrete plate; (b) power transmissibility ($\gamma$); and (c) force transmitted to floor through each contact point against disturbing frequency (natural frequency: 15 Hz).

parts of all mobilities for the four contact points have similar shapes, only one floor mobility, $Y_{11}$, is shown in Figure 6.11(a). The floor mobility is inconsiderable at most frequencies (except between 48.5 and 50.5 Hz). A peak mobility value is observed at frequencies between 48.5 and 50.5 Hz, approximating the resonant frequency of the floor. If the machine mass, $M$, is 2,000 kg, $\alpha^2$ is 1.8 for the rectangular steel machine model with $L$ = 1.485 m, $l$ = 1.185 m, and $H$ = 0.32 m; the steel density is $\rho s = 7.85 \times 10^3$ kg·m$^{-3}$; and the natural frequency is 15 Hz. The power transmissibility, $\gamma$ (Eq. 4.24), against distributing frequency for a rectangular steel machine is shown in Figure 6.11(b).

Figure 6.11(b) shows that at most frequencies (except between 48.5 and 50.5 Hz), the power transmissibility, $\gamma$, is less than 1 and decreases with increasing frequencies starting from 50.5 Hz. Figure 6.12 shows that the forces transmitted to the floor through four contact points considerably decrease after the installation of spring isolators at frequencies in the range

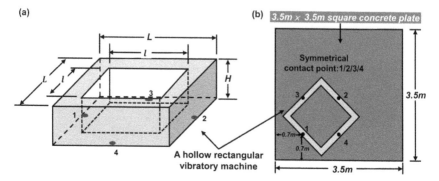

*Figure 6.10* (a) Three-dimensional view of hollow vibratory machine with external dimensions of $L \times L \times H$ and internal dimensions of $l \times l \times H$; (b) Plan view of modelled situation of machine placed asymmetrically on a simply-simply-simply-simply square concrete plate with four symmetrical supports at contact points 1, 2, 3, and 4.

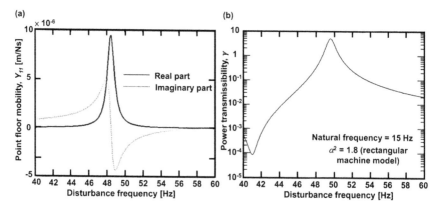

*Figure 6.11* Plots of (a) point floor mobility ($Y_{11}$) and (b) power transmissibility ($\gamma$) against disturbance frequency at natural frequency (15 Hz) for $\alpha^2 = 1.8$ (rectangular machine model).

40–60 Hz (approximately 1/6 of that without a spring isolator). However, interestingly, in Figure 6.11(b), the power transmissibility, $\gamma$, considerably exceeds 1 (i.e., approximately 3–4) at frequencies near 50 Hz. This means that the total active power transmitted to the floor with spring isolators is three to four times greater than that without spring isolators. Force transmissibility is based on the force transmitted to the floor through the contact point, thus providing information regarding this force. In contrast, power transmissibility provides information pertaining to the actual total power transmitted to the floor through all contact points. At a forcing frequency of 50 Hz and natural frequency of 15 Hz, the dynamic forces transmitted to the floor through contact points 1, 2, 3, and 4 decrease from 0.573 + 0.016j (magnitude 0.573) to –0.038 – 0.015j (magnitude: 0.041); from 0.315 + 0.007j (magnitude 0.315) to –0.064 – 0.043j (magnitude: 0.077); from –0.365–0.030j (magnitude: 0.366) to –0.048 – 0.025j (magnitude: 0.054); and from –0.365 – 0.030j (magnitude: 0.366) to –0.048 – 0.025j (magnitude: 0.054), respectively. Evidently, the magnitudes of the forces transmitted to the floor through each contact point decrease and have different phases after the installation of the spring isolator. Owing to the interaction among the four contact points (i.e., the phase relationships among the four transmitted forces), force cancellations or reinforcing forces are required among the four transmitted forces. The total power transmitted to the floor is therefore the result of the four transmitted forces. The result of the total active power transmitted to the floor through the four contact points at frequencies approximating the resonant frequency of the floor shown in Figure 6.11(b) is found to have an opposite effect. Different from the forces transmitted to the floor, the total active power transmitted increases at a forcing frequency of 50 Hz after the installation of the spring isolators. Nevertheless, Figures 6.11(b) and 6.12 show that both the force and total

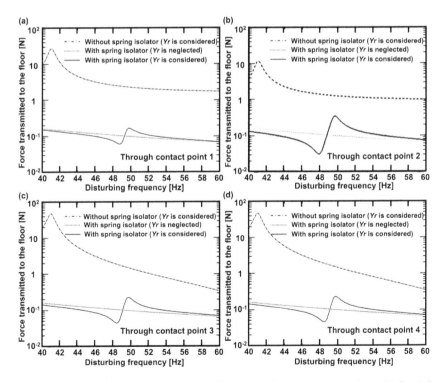

*Figure 6.12* Plots of forces transmitted to floor through contact points (a) 1; (b) 2; (c) 3; and (d) 4.

active power transmitted to the floor decrease at most frequencies (except at frequencies near the resonant frequency of the floor) after the installation of the spring isolators. Therefore, different from force transmissibility, power transmissibility provides information regarding the actual power transmitted to the floor and has engineering importance. In this example, force transmissibility indicates that the spring isolator operates well at all frequencies, including forcing frequencies near 50 Hz; however, this conclusion is inaccurate.

## 6.5 PREDICTION OF PSYCHOLOGICAL RESPONSES

An air-conditioning ventilation system is an essential component of buildings to maintain a good indoor environmental quality. Satisfactory ventilation system design must minimise the negative impact of noise on occupants. This case study endeavours to demonstrate how the holistic psychoacoustic approach, discussed in Chapter 5, can aid in the investigation of the multi-dimensional psychological impacts of ventilation noise on occupants. The objective characteristics of different ventilation noise are assessed using a list of traditional acoustic and psychoacoustic metrics. The noise impact on the perception of occupants and degree of attention are assessed using the psychoacoustics perception scale (PPS) [4] and letter cancellation task (LCT), respectively. The statistical analysis results of the case study indicate the means for efficiently quantifying the psychological impacts of ventilation noise on the general judgement (*Evaluation, E*) and magnitude sensitivity (*Potency, P*) of occupants. Moreover, the results show the approach for determining the sensation of the temporal and spectral compositions (*Activity, A*) [5] of the ventilation noise with the aid of the PPS. They also identify the technique for analysing the association between the objective characteristics of ventilation noise and perceptual effects on other psychological impacts, such as annoyance, discomfort, unpleasantness, and unacceptability.

### 6.5.1 Objective characteristics of ventilation noise

In the case study, 12 ventilation noise soundtracks were recorded in a university classroom with dimensions of 8.4 m (*L*) × 7.4 m (*W*) × 3 m (*H*) (Figure 6.13). All classroom windows and doors were closed. To minimise the impact of other environmental noise instead of ventilation noise, only one experimenter was inside the classroom to control the advanced hand-held analyser (Type 2270; Bruel & Kjaer (B&K), Naerum, Denmark) for soundtrack recording. Measurement points (*MPs*) 1 and 4, 2 and 5, and 3 and 6 are labelled as rows 1, 2, and 3 for the experiment, respectively. The flow direction in the air duct is from rows 1 to 3. Ventilation and cooling modes could be controlled by the control panel of the centralised ventilation

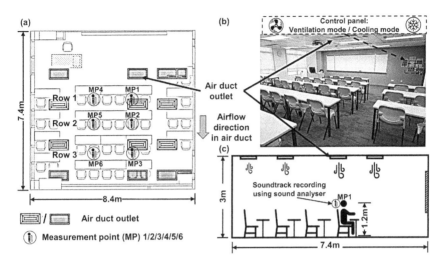

*Figure 6.13* (a) Floor plan of university classroom used for psychoacoustic study showing locations of *MPs* and air duct outlets; (b) on-site photo of university classroom; and (c) schematic of sound recording measurement using sound analyser.

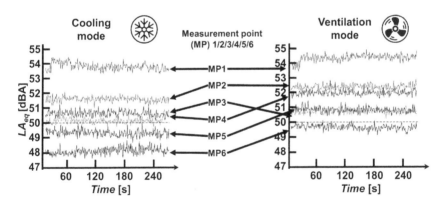

*Figure 6.14* First 4 min of time-varying data of A-weighted $L_{Aeq}$ of 12 soundtracks recorded at 6 *MPs* in two operation modes.

system of the classroom (Figure 6.13(b)). For each soundtrack recording, the B&K Type 2270 sound analyser was placed at a height of 1.2 m (average seated ear height) from the ground surface (Figure 6.13(c)). The duration of each recording was 10 min. The first 4 min of the time-varying data of the A-weighted $L_{Aeq}$ of recordings at the *MPs* is shown in Figure 6.14. In general, the value of $L_{Aeq}$ of noise in the ventilation mode exceeded than that

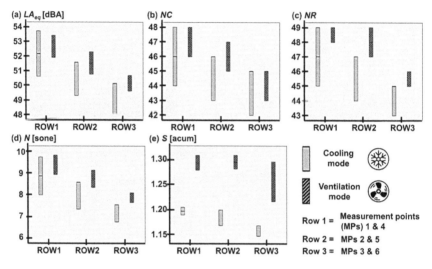

*Figure 6.15* Multiple boxplots of objective characteristics of ventilation noise: (a) A-weighted $L_{Aeq}$; (b) NC; (c) NR; (d) N; and (e) S in different operation modes (cooling and ventilation) and rows.

in the cooling mode. The $L_{Aeq}$ value decreased along the airflow direction (from rows 1 to 3). As for the noise under the air duct outlet (*MPs* 1, 2, and 3), the $L_{Aeq}$ value exceeded those at *MPs* 4, 5, and 6.

Consistent results were found among the energy content-related acoustic metrics, such as $L_{Aeq}$, noise criteria (NC), noise rating (NR), and N (Figure 6.15(a)–(d), respectively). The measurement of the spectral content-related psychoacoustic metric, i.e., sharpness (S), showed that the change in spectral content was smaller along the airflow direction compared with that in operational modes (Figure 6.15(e)). A relatively high low-frequency component can be found for the noise in the cooling mode compared with that in the ventilation mode.

## 6.5.2 Psychological responses to ventilation noise

The jury listening test is a method for the subjective assessment of a subject's responses to recorded soundtracks. The tests involved the playback of the soundtracks of ventilation noise in an anechoic chamber (Figure 6.16). The omnidirectional source, LS02, an amplified dodecahedron loudspeaker for playing the recorded soundtracks was placed 2 m away from the participants. A 1-kHz pure tone sound at 60 dB was used as the calibration signal. A B&K Type 2270 sound analyser was used to ensure that the magnitude of the calibration signal heard at the location of the participants was 60 dB.

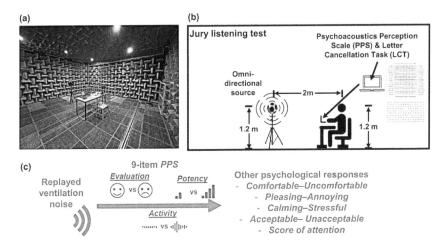

*Figure 6.16* (a) On-site photograph of anechoic chamber and schematic of jury listening test using PPS and LCT.

### 6.5.2.1 Psychoacoustics perception scale

The measurement of multidimensional psychological impacts was a self-administered questionnaire combining the PPS and other negative perceptions of sounds. Soundtracks extracted for 1 min were played during the jury listening test. The participants were asked to rate their responses to the nine psychological impacts proposed in the PPS (*E* dimension: 'Quiet–Noisy', 'Relaxed–Tense', and 'Pleasant–Unpleasant'; *P* dimension: 'Weak–Strong', 'Quiet–Loud', and 'Light–Heavy'; and *A* dimension: 'Low–High', 'Dull–Sharp', and 'Deep–Metallic'; and *A2* and *A3*). The three other negative perceptions of sounds were 'Comfortable–Uncomfortable', 'Pleasing–Annoying', 'Calming–Stressful', and 'Acceptable–Unacceptable'. All the questions were on a semantic differential scale formed by a bipolar semantic pair with opposite meanings. For example, the semantic pair 'Quiet–Noisy' has seven levels. The perceptual responses, 'Extremely Quiet', 'Quite Quiet', 'Slightly Quiet', 'Equally Quiet/Noisy', 'Slightly Noisy', 'Quite Noisy', and 'Extremely Noisy', were rated from –3 to +3.

### 6.5.2.2 Letter cancellation task

The LCT is a paper-and-pencil test widely used to measure the attention of participants rapidly. A table of 23 random letters per row and 10 columns was provided to the participants in each jury listening test on a 1-min soundtrack. The participants were required to cancel all 'A' letters (a total of 40) on the table. The percentage of correct cancellations of 'A' letters was the attention score.

### 6.5.3 Statistical analysis

All data in the statistical analysis were coded and analysed using the commercial package SPSS (version 23.0; IBM Corp., Armonk, NY, USA). All significance levels in the tests were set at 0.05. The Mann–Whitney $U$ test was used to compare the differences among the psychological impacts of ventilation noise on participants in different operation modes. Moreover, Spearman's rank correlation was applied to test the correlations between the objective characteristics of ventilation noise and multidimensional psychological impacts or among the multidimensional psychological impacts.

#### 6.5.3.1 Descriptive analysis of data

A total of 68 jury listening tests were conducted. Half of the tests involved ventilation noise in ventilation mode, whereas the other half involved noise in the cooling mode. Of the total participants, 64.7% were male, and 35.3% were female. In general, the mean values of the negative psychological impact ratings were higher for ventilation noise (Figure 6.17). The participants felt significant tension due to the noise in the ventilation mode (mean ± standard derivation = 0.68 ± 1.04) than in the cooling mode (−0.06 ± 1.32). The results of objective characteristics in terms of $L_{Aeq}$, NC, NR, N, and S of ventilation noise showed that both the energy and spectral contents of the noise could be affected by the changes in the operation mode. The perceptual influence of the change in objective properties was further assessed using the PPS. Higher scores in the E, P, and A dimensions indicate worse general judgement, higher sensitivity to noise magnitude, and greater sensation to high frequencies of sounds, respectively. The measurement results of the PPS further quantified the perceptual impacts on occupants owing to the change in the operation mode. The noise with high magnitude and frequency in the ventilation mode also affected the occupants' perceptions of noise, especially for the 'Tense' perception ($p < 0.05$).

#### 6.5.3.2 Data correlation analysis

In the analysis of psychological impacts on the E dimension, the negative impact on 'Quiet–Noisy' was found to be significantly correlated with all the acoustic and psychoacoustic metrics ($ps < 0.05$; Figure 6.18(a)). In the P dimension, all related negative impacts were found to be significantly correlated with the physical and psychoacoustic metrics of noise magnitude ($ps < 0.05$). However, there was no significant association between the psychological impacts on the A dimension and objective metrics.

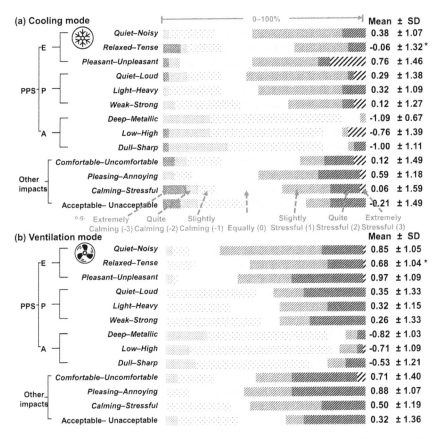

*Figure 6.17* Psychological impacts of ventilation noise in different operation modes: (a) cooling mode and (b) ventilation mode (* $p < 0.05$ in Mann–Whitney $U$ tests of variables in two operation modes).

In the analysis between the PPS and other psychological impacts, 'Comfortable–Uncomfortable', 'Pleasing–Annoying', 'Calming–Stressful', and 'Acceptable–Unacceptable' were found to be significantly correlated with the psychological impacts on the $E$ and $P$ dimensions ($ps < 0.001$; Figure 6.18(b)). The Spearman's rank correlation coefficients (0.65–0.80) between the other negative impacts and $E$ dimension exceeded those of the $P$ dimension (0.46–0.61).

The ability of PPS to capture hypothetical perceptual dimensions ($E$, $P$, and $A$) was further verified using the Spearman's rank correlation test results. Evidently, the questions regarding the $P$ dimension can quantify the increments of noise magnitude in terms of $L_{Aeq}$, NC, NR, and $N$ ($ps < 0.05$). The questions regarding $E$ can quantify the general sound quality

(a)

| | LAeq | NC | NR | N | S |
|---|---|---|---|---|---|
| Quiet–Noisy | 0.37** | 0.31* | 0.39*** | 0.38** | 0.34** |
| Relaxed–Tense | 0.23 | 0.14 | 0.27* | 0.27* | 0.35** |
| Pleasant–Unpleasant | 0.10 | 0.03 | 0.12 | 0.13 | 0.15 |
| Quiet–Loud | 0.37** | 0.33** | 0.36** | 0.34** | 0.11 |
| Light–Heavy | 0.27* | 0.20 | 0.25* | 0.27* | 0.12 |
| Weak–Strong | 0.37** | 0.32** | 0.35** | 0.36** | 0.15 |
| Deep–Metallic | -0.03 | -0.01 | -0.01 | -0.02 | 0.02 |
| Low–High | -0.16 | -0.18 | -0.14 | -0.14 | 0.02 |
| Dull–Sharp | -0.05 | -0.06 | -0.01 | -0.02 | 0.14 |
| Comfortable–Uncomfortable | 0.23 | 0.18 | 0.25* | .26* | 0.21 |
| Pleasing–Annoying | 0.20 | 0.14 | 0.21 | 0.24 | 0.21 |
| Calming–Stressful | 0.25* | 0.18 | 0.26* | 0.28* | 0.21 |
| Acceptable–Unacceptable | 0.23 | 0.14 | 0.23 | 0.26* | 0.25* |

(b)

| | Comfortable–Uncomfortable | Pleasing–Annoying | Calming–Stressful | Acceptable–Unacceptable |
|---|---|---|---|---|
| Quiet–Noisy | 0.65*** | 0.64*** | 0.69*** | 0.64*** |
| Relaxed–Tense | 0.74*** | 0.69*** | 0.80*** | 0.72*** |
| Pleasant–Unpleasant | 0.65*** | 0.69*** | 0.64*** | 0.73*** |
| Quiet–Loud | 0.56*** | 0.51*** | 0.61*** | 0.52*** |
| Light–Heavy | 0.47*** | 0.51*** | 0.52*** | 0.52*** |
| Weak–Strong | 0.46*** | 0.49*** | 0.55*** | 0.53*** |
| Deep–Metallic | -0.03 | 0.09 | -0.02 | 0.06 |
| Low–High | -0.02 | 0.08 | -0.10 | 0.08 |
| Dull–Sharp | 0.13 | 0.21 | 0.14 | 0.20 |

Spearman's rank correlation coefficient ($\rho$)

-1    0    1

*Figure 6.18* Spearman's rank correlation coefficients (a) between objective characteristics of ventilation noise and psychological impacts and (b) among psychological impacts.

of ventilation noise with the change in both energy and spectral contents ($L_{Aeq}$, NC, NR, N, and S ($ps < 0.05$)). Because only ventilation noise was considered in this study, the participants could not differentiate the changes in temporal and spectral contents among the recorded sounds.

Furthermore, the scores of responses to the questions in the PPS were found to be excellent subjective indicators of other negative psychological impacts. The feelings of participants associated with 'Uncomfortable', 'Annoying', 'Stressful', and 'Unacceptable' were significantly correlated with their responses to the $E$ and $P$ dimensions. A high Spearman's rank correlation coefficient of responses to the questions in the $E$ dimension indicated that the improvement in the general sound quality was more important than a mere reduction in noise magnitude. Excellent sound quality is not mere quiescence. Moreover, when PPS assessment is not implemented, the environmental impact of the annoyance of participants due to changes in the objective environment is not observed (i.e., the non-significant results between objective characteristics and psychological impacts shown in Figure 16.8(a)). The foregoing demonstrates the necessity for PPS assessment in the holistic psychoacoustic study of environmental noise impacts.

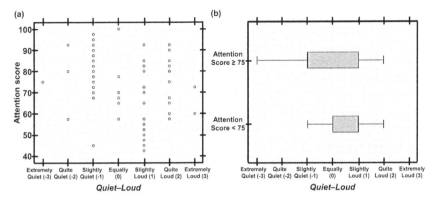

*Figure 6.19* (a) Scatter plot of participants' attention score and their responses to 'Quiet–Loud' perception and (b) boxplots of responses to 'Quiet–Loud' perception of participants with attention scores ≥ 75 and < 75.

For the analysis of the attention score from the LCT, 'Quiet–Loud' was found to be significantly correlated with the attention score of participants (Spearman's rank correlation coefficient, $\rho = 0.29$ ($p < 0.05$); Figure 6.19(a)). The participants had a higher attention score if they felt that the ventilation noise was quieter. After further grouping the participants into two (one group with an attention score of ≥75 and another group with an attention score of <75), the group with a high attention score was found to have positive perception ('Slightly Quiet' (Mean = –0.094)) of ventilation noise. Those with a low score had negative perception ('Slightly Noisy' (Mean = 0.69 and $p < 0.05$ in the Mann–Whitney $U$ test of attention score between two participant groups); Figure 6.19(b)).

## 6.5.4 Significance of results

Successful and sustainable noise control requires knowledge of the multi-dimensional psychological impacts of noise on occupants. The different applications of PPS provide reliable points of comparison between different acoustic and psychoacoustic studies in sustainable noise control designs. The PPS successfully recorded negative perceptual impacts in terms of dimensions other than noise magnitude. It is a more advanced approach compared with the traditional method, which only focuses on the participants' perception of noise loudness. Statistical analysis results show that the evaluation of perceptual responses using the PPS is key to understanding the participants' psychological responses (i.e., 'Comfortable–Uncomfortable', 'Pleasing–Annoying', 'Calming–Stressful', and 'Acceptable–Unacceptable') to ventilation noise. The negative responses of participants were not only influenced by the loudness sensation but also by the general judgement of noise. Furthermore, through the PPS, the use of psychoacoustic metrics $N$ and $S$ to understand environmental noise impacts is demonstrated. The use of PPS can also aid in specifying the mechanism of environmental perceptual impacts on the attention of participants. These findings provide knowledge on creating a healthy and pleasant air-conditioned building environment by appropriately mitigating the multidimensional negative psychological impacts on occupants.

## 6.6 SUMMARY

The case studies presented in this chapter are step-by-step examples to demonstrate how the knowledge presented in previous chapters can be applied to the prediction of duct transmission noise, flow-generated noise, vibration isolation performance, and human psychoacoustic responses. Such applications enable the resolution of real-life ventilation noise problems.

## References

1. C.M. Mak, J. Wu, C. Ye, and J. Yang, 2009 *The Journal of the Acoustical Society of America* 125, pp.3756–3765, Flow noise from spoilers in ducts.
2. C.M. Mak and J. Su, 2002 *Applied Acoustics* 63, pp.1281–1299, A power transmissibility method for assessing the performance of vibration isolation of building services equipment.
3. L. Cremer, M. Heckl, 1973 *Structure-borne sound*. Springer-Verlag: New York.
4. K.W. Ma, C.M. Mak, and H.M. Wong, 2020 *Journal of Building Engineering* 29, p.101177, Development of a subjective scale for sound quality assessments in building acoustics.
5. K.W. Ma, H.M. Wong, and C.M. Mak, 2018 *Building and Environment* 133, pp.123–150, A systematic review of human perceptual dimensions of sound: Meta-analysis of semantic differential method applications to indoor and outdoor sounds.

# Index

For Product Safety Concerns and Information please contact our EU
representative GPSR@taylorandfrancis.com
Taylor & Francis Verlag GmbH, Kaufingerstraße 24, 80331 München, Germany

www.ingramcontent.com/pod-product-compliance
Ingram Content Group UK Ltd.
Pitfield, Milton Keynes, MK11 3LW, UK
UKHW021122180425
457613UK00005B/190